IMPROVING METRICS for the Department of Defense Cooperative Threat Reduction Program

Committee on Improving Metrics for the Department of Defense
Cooperative Threat Reduction Program

Committee on International Security and Arms Control

NATIONAL ACADEMY OF SCIENCES
THE NATIONAL ACADEMIES

THE NATIONAL ACADEMIES PRESS
Washington, D.C.
www.nap.edu

THE NATIONAL ACADEMIES PRESS 500 Fifth Street, N.W. Washington, DC 20001

NOTICE: The project that is the subject of this report was approved by the Governing Board of the National Research Council, whose members are drawn from the councils of the National Academy of Sciences, the National Academy of Engineering, and the Institute of Medicine. The members of the committee responsible for the report were chosen for their special competences and with regard for appropriate balance.

This study was supported by Contract/Grant No. HDRTA1-10-C00058 between the National Academy of Sciences and Department of Defense. Any opinions, findings, conclusions, or recommendations expressed in this publication are those of the author(s) and do not necessarily reflect the views of the organizations or agencies that provided support for the project.

International Standard Book Number 13: 978-0-309-22255-6
International Standard Book Number 10: 0-309-22255-9

Limited copies of this report are available from the Committee on International Security and Arms Control, 500 5th Street, N.W., Washington, DC 20001, (202) 334-2811, cisac@nas.edu.

Additional copies of this report are available from the National Academies Press, 500 Fifth Street, N.W., Lockbox 285, Washington, DC 20055; (800) 624-6242 or (202) 334-3313 (in the Washington metropolitan area); Internet, http://www.nap.edu.

Printed in the United States of America

THE NATIONAL ACADEMIES

Advisers to the Nation on Science, Engineering, and Medicine

The **National Academy of Sciences** is a private, nonprofit, self-perpetuating society of distinguished scholars engaged in scientific and engineering research, dedicated to the furtherance of science and technology and to their use for the general welfare. Upon the authority of the charter granted to it by the Congress in 1863, the Academy has a mandate that requires it to advise the federal government on scientific and technical matters. Dr. Ralph J. Cicerone is president of the National Academy of Sciences.

The **National Academy of Engineering** was established in 1964, under the charter of the National Academy of Sciences, as a parallel organization of outstanding engineers. It is autonomous in its administration and in the selection of its members, sharing with the National Academy of Sciences the responsibility for advising the federal government. The National Academy of Engineering also sponsors engineering programs aimed at meeting national needs, encourages education and research, and recognizes the superior achievements of engineers. Dr. Charles M. Vest is president of the National Academy of Engineering.

The **Institute of Medicine** was established in 1970 by the National Academy of Sciences to secure the services of eminent members of appropriate professions in the examination of policy matters pertaining to the health of the public. The Institute acts under the responsibility given to the National Academy of Sciences by its congressional charter to be an adviser to the federal government and, upon its own initiative, to identify issues of medical care, research, and education. Dr. Harvey V. Fineberg is president of the Institute of Medicine.

The **National Research Council** was organized by the National Academy of Sciences in 1916 to associate the broad community of science and technology with the Academy's purposes of furthering knowledge and advising the federal government. Functioning in accordance with general policies determined by the Academy, the Council has become the principal operating agency of both the National Academy of Sciences and the National Academy of Engineering in providing services to the government, the public, and the scientific and engineering communities. The Council is administered jointly by both Academies and the Institute of Medicine. Dr. Ralph J. Cicerone and Dr. Charles M. Vest are chair and vice chair, respectively, of the National Research Council.

www.national-academies.org

COMMITTEE ON IMPROVING METRICS FOR THE DEPARTMENT OF DEFENSE COOPERATIVE THREAT REDUCTION PROGRAM

Jay C. Davis, *Chair,* Hertz Foundation, Livermore, California
George W. Anderson, Jr., Lawrence Livermore National Laboratory, California
Steven J. Gitomer, National Science Foundation, Santa Fe, New Mexico
Mary Alice Hayward, Areva, Inc., Bethesda, Maryland
Mark F. Mullen, Los Alamos National Laboratory, Los Alamos, New Mexico
Gregory S. Parnell, United States Military Academy, West Point, New York
Kim K. Savit, SAIC, McLean, Virginia; University of Denver, Colorado
Nicolas Van de Walle, Cornell University, Ithaca, New York

National Research Council Staff

Micah D. Lowenthal, Study Director (November 2010 to completion)
Anne M. Harrington, Study Director (through October 2010)
Glenn Schweitzer, Director, Program on Central Europe and Eurasia
Rita S. Guenther, Program Officer
La'Faye Lewis-Oliver, Administrative Coordinator

PREFACE AND ACKNOWLEDGMENTS

The Cooperative Threat Reduction (CTR) Program was created in 1991 as a set of support activities assisting the Former Soviet Union states in securing and eliminating strategic nuclear weapons and the materials used to create them. The Program evolved as needs and opportunities changed: Efforts to address biological and chemical threats were added, as was a program aimed at preventing cross-border smuggling of weapons of mass destruction. (More detail can be found in the Background section of Chapter 1.)

CTR has traveled through uncharted territory since its inception, and both the United States and its partners have taken bold steps resulting in progress unimagined in initial years. In a few cases, the program made bold missteps and became mired in projects that foundered in mistrust, misaligned objectives, and political knots. Over the years, much of the debate about CTR on Capitol Hill has concerned the effective use of funds, when the partners would take full responsibility for the efforts, and how progress, impact, and effectiveness should be measured.

CTR is now 20 years old and many of the same questions remain. Increasingly, there is a push for metrics to ensure proper guidance and execution of the Program. Congress directed the Department of Defense (DoD) to request this review of DoD's CTR metrics development.

The National Research Council appointed an ad hoc committee to conduct the study. The committee held several public information-gathering meetings, which included presentations by representatives of DoD CTR. We would like to thank the management and staff of the DoD CTR Program; Luke Kluchko and Shawn Anderson of the Defense Threat Reduction Office in Kyiv, Ukraine; Ambassador Bonnie Jenkins; Assistant Secretary of Defense Andrew Weber; Andrew Hood of the Science and Technology Council of Ukraine; Dan Rosenbaum and Ben Klay of the Office of Management and Budget; Joseph Christoff at the Government Accountability Office; Tom Karako of the House Armed Services Committee staff; Julie Unmacht, formerly on the House Armed Services Committee staff; and Assistant Secretary of Defense Madelyn Creedon, then of the Senate Armed Services Committee staff.

The DoD CTR Metrics Report was issued in late 2010 and the first implementation of new metrics was done in early 2011. Only preliminary results were available during the information-gathering stage of the study. Observations from DoD's presentations on implementation of the metrics were incorporated into the evaluation in this report. In addition, the committee reviewed existing literature on and experiences with the development of metrics in complex, difficult to measure contexts. Finally, to obtain an understanding of the realistic opportunities and challenges associated with DoD CTR programs as they are currently implemented, a small group of committee members traveled to Ukraine to conduct site visits and meet with Ukrainian officials who implement projects (land and maritime border security, human and animal health) under the CTR Program. This on-the-ground view provided useful insights as the committee finalized its findings and recommendations. This information-gathering trip was not, however, a review of the CTR Program in Ukraine, nor was Ukraine considered by the committee to be representative of other countries. So while some examples in the report are drawn from that trip, the committee has not provided a full trip report, although an itinerary is included in Appendix D.

This report has been reviewed in draft form by individuals chosen for their diverse perspectives and technical expertise, in accordance with procedures approved by the National Academies' Report Review Committee. The purpose of this independent review is to provide candid and critical comments that will assist the institution in making its published report as sound as possible and to ensure that the report meets institutional standards for objectivity, evidence, and responsiveness to the study charge. The review comments and draft manuscript remain confidential to protect the integrity of the process.

We wish to thank the following individuals for their review of this report: Jayson Ahern, U.S. Customs and Border Protection; Dave Franz, Midwest Research Institute; Gerald Epstein, American Association for the Advancement of Science; Sally Katzen, Podesta Group; L. Robin Keller, University of California at Irvine; Susan Koch, U. S. Department of State (Retired); Devra Moehler, University of Pennsylvania; Mary Beth Nikitin, Congressional Research Service; Thomas Pickering, Hills and Company; and Jeffery Richardson, Center for International Security and Cooperation.

Although the reviewers listed above have provided many constructive comments and suggestions, they were not asked to endorse the conclusions or recommendations, nor did they see the final draft of the report before its release. The review of this report was overseen by Chris Whipple of Environ Corporation and Mona Dreicer of Lawrence Livermore National Laboratory. Appointed by the National Academies, they were responsible for making certain that an independent examination of this report was carried out in accordance with institutional procedures and that all review comments were carefully considered. Responsibility for the final content of this report rests entirely with the authoring committee and the institution.

As the DoD CTR Program enters its third decade and seeks to address new security and threat reduction challenges in a variety of countries and contexts, it is particularly appropriate for DoD, together with its partners, to update its approach to assessing programmatic impact and effectiveness for program management as well as for overall strategic planning. These metrics must be linked to concisely-stated threat reduction objectives, be jointly developed with partner countries, be prioritized, leverage the experience and insights of other agencies, and reflect change over time, including explicitly-defined sustainability metrics. Such metrics will provide a more informed basis upon which to make critical decisions for DoD CTR programs into the future. While the metrics task is challenging, the committee believes that the recommendations outlined here provide a solid and useful approach for all those who must meet the on-going need for effective and efficient metrics at all levels of the DoD CTR Program.

Jay Davis
Chair

CONTENTS

SYNOPSIS

This National Academy of Sciences (NAS) study, requested under Congressional mandate, evaluates the "metrics to measure the impact and effectiveness of activities of the Cooperative Threat Reduction [CTR] Program of the Department of Defense [DoD]" (Section 1304, P.L. 111-84) described in a DoD September 2010 report (DoD, 2010; hereinafter the DoD Metrics Report). The NAS study committee found that the DoD Metrics Report provides reasonable metrics for some activities of the CTR Program which consolidate and eliminate weapons of mass destruction and weapons materials, and provides a solid starting point for developing metrics for newer, expanded capacity-building efforts under the CTR Program. The committee also identified shortcomings in the DoD Metrics Report and here provides recommendations to enhance DoD's development and use of metrics for the CTR Program.

Improving Metrics

The committee has several recommendations to DoD on how it can improve its metrics for the CTR programs and better communicate its approach to metrics.

1. For each program described in the DoD Metrics Report, DoD should include a concise statement of its objectives and of how the program is intended to reduce threat or risk.
2. Objectives for projects and the overall CTR Program in a partner country are developed jointly between the United States and the partner country. An agreed set of metrics should also be built into projects from the outset. They may change, but the parties responsible for the projects should know at any given time the metrics that will be used to measure impact and effectiveness.
3. The committee judges that using a consistent framework to prioritize and refine metrics within a program would help DoD and other CTR decision makers. Using such a framework, DoD can identify the highest priority metrics, ensuring that the metrics are useable and useful, and allow decision makers to feed results back into the overall CTR processes of determining objectives and budgets. Any of several decision-making or prioritizing frameworks would be feasible, including the decision analysis technique of swing-weight analysis and the DoD capabilities based planning process.
4. DoD plans to leverage other U.S. Government agencies' experience, capabilities, and assets as CTR expands to new countries and as it continues existing programs. DoD also needs to communicate, coordinate, and cooperate with relevant agencies.
5. DoD's metrics and planning process should more explicitly factor in both planned and unplanned change over time. During the phases of active DoD involvement in a CTR project and afterward during sustainment, which is its own stage requiring resources (budgets, equipment, and trained people), clearer planning for how changes and metrics results will feed into decision making will make the metrics more credible and useful for both DoD and the partner country.
6. Capacity building programs need independent evaluation of how the capabilities being built perform in action. This can be accomplished by several means, ranging from expert observations of routine operations to comprehensive exercises that test the full scope of

capabilities. The level of effort can be tailored to the scope of the program, its resources, and its relative importance. DoD and its partners should build such independent evaluation into each project. The Defense Security Cooperation Program might be a good model for how to proceed.

SUMMARY

Through the Cooperative Threat Reduction (CTR) Program, the United States works with partner countries to address threats of mutual concern that could manifest in, transit through, or emanate from their territory.[1] Congress directed the Secretary of Defense to "develop and implement metrics to measure the impact and effectiveness of activities of the CTR Program of the Department of Defense [DoD] to address threats arising from the proliferation of chemical, nuclear, and biological weapons and weapons-related materials, technologies, and expertise" (Section 1304, P.L. 111-84, see Appendix A). The Secretary completed a report describing DoD's metrics for the CTR Program (DoD, 2010; here called the DoD Metrics Report, see Appendix B) in September 2010 and, as required in the same law, contracted with the National Academy of Sciences to review the metrics DoD developed and identify possible additional or alternative metrics, if necessary. This report provides that review and advice.

What is a metric? As implied in the language cited above, a metric is an evidence-based tool that measures impact and effectiveness, which can be defined in terms of the performance of a program or a project with respect to its objectives. Metrics alone cannot ensure that the best options have been identified or are being implemented, nor by themselves can they tell managers and decision makers why progress is or is not being made (see What Metrics Cannot Do, in Chapter 3), but they can be helpful in establishing reference points and informing decision makers of whether the efforts are bearing fruit.

The CTR Program's early focus was primarily on dismantlement and nonproliferation of weapons of mass destruction (WMD), WMD materials, and WMD expertise in the former Soviet Union. The Program has evolved both geographically and topically to include work aimed at improving partner nations' abilities to deter, detect, and respond to emerging WMD threats, and the Program is envisioned to shift further in this direction. In sum, the Program has shifted from dealing with specific sources of known risk—weapons, weapons materials, and expertise that had been associated with actual WMD programs—to dealing with potential sources of future risk. This is particularly true of the CTR Cooperative Biological Engagement Program (CBEP). CBEP's work resides in a gray mission space that has elements that overlap with public health activities. Impact, effectiveness, and success are difficult to measure in such efforts—there are no simple parallel metrics to counting the number of delivery vehicles destroyed or the fraction of weapons-useable nuclear material secured or eliminated that can effectively measure the impact of any complex capacity building program, such as CBEP. It is complex and challenging to develop metrics for the partner's capabilities and the personal and institutional relationships established, which are the main products of a successful capacity-building program. As a result, the committee concludes that it is possible to successfully accomplish what is easily measurable and still fail in the engagement.

[1] The CTR Program began in 1991 as a means of assisting the former Soviet Union and later additional countries with strategic offensive arms elimination; nuclear warhead dismantlement; nuclear weapons storage security; chemical weapons destruction; biological weapons proliferation prevention; reactor core conversions; nuclear material protection, control and accounting; export control initiatives; defense conversion; as well as other projects. Across the U.S. Government, CTR projects are administered by DoD, the Department of Energy, the Department of Commerce, and the Department of State. This report addresses only DoD programs and projects. The DoD CTR Program was incorporated into the Defense Threat Reduction Agency (DTRA) when the agency was established in 1998.

ASSESSMENT OF THE DOD METRICS REPORT

The committee evaluated the DoD CTR metrics described in the DoD Metrics Report based on whether the metrics provide decision makers the essential information to manage the effectiveness and impact of CTR programs. Most of the CTR programs try to develop a sustaining capability of some sort through a set of projects implemented with partner countries on their territory. The DoD Metrics Report contains reasonable metrics for the CTR programs that consolidate and eliminate weapons and weapons materials,[2] and contains a solid starting point for developing metrics for the newer capacity-building programs. For meaningful evaluation, the committee assessed that DoD must (1) state the objectives of the program and the projects (i.e., the goals of the actual activities); (2) identify the capabilities it is trying to develop or maintain; (3) link those capabilities to metrics; (4) ensure that the metrics reflect program effectiveness and impact; and (5) plan for and measure sustainment.[3] It is generally good practice for the program to establish minimum performance levels (performance that must be achieved for the project to not fail) and aspirational goals (the desired performance above the minimum) for each metric.[4]

- Overall, the DoD Metrics Report describes CTR's highest-level objectives and difficulties in developing metrics clearly and succinctly in the introductory section. However, the report does not connect these objectives to threats for the capacity-building programs. This is not to say that there is no connection, nor that the explanation does not appear in other documents,[5] just that the DoD Metrics Report does not describe the connection. For example, describing the connection would clarify how building the capacity to better track respiratory disease in East Africa reduces threats to U.S. national security. The lack of a concise statement of the objectives of each program and how the actions planned under the program are intended to reduce threat or risk is a deficiency that makes the Report less effective for communicating with people outside of the program and internally makes development and refinement of metrics more difficult.
- CTR programs are meant to be partnerships and work best when they are carried out as partnerships with other countries (hence, *cooperative* threat reduction). This includes joint development of the objectives with the partner countries. The DoD Metrics Report does not make clear whether and how partner countries participate in determining objectives and metrics, or the partner country's role in measurement. The report reads as if metrics are U.S. measurements of the partner's progress toward U.S. goals rather than the impact and effectiveness of the program measured with jointly agreed metrics.

---

[2] The committee did not address the Strategic Offensive Arms Elimination Program, which is not discussed in the DoD Metrics Report because DoD plans to use the long-standing "Nunn-Lugar Scorecard" metrics for that program.

[3] In its report, DoD seems to use the term sustainability to refer to both the ability to sustain the program, security state, or other improvements of the CTR programs, and to the actual result or act of sustaining them. The committee refers to the former as sustainability and the latter as sustainment.

[4] In capabilities based planning, these are referred to as threshold values and objective values. Because the term "objective" is used in so many different ways, the committee has not adopted this usage.

[5] See, for example, Nacht (2009).

- DoD did not use a consistent framework for developing and articulating program objectives, capacities, and metrics and did not prioritize among its metrics. As a consequence, the DoD Metrics Report mixes project management measures with higher-level program performance metrics for some of the CTR programs and weights equally metrics that are critically important and others are not.
- DoD plans to leverage other U.S. Government agencies' experience, capabilities, and assets as CTR expands to new countries and as it continues existing programs. DoD also needs to communicate, coordinate, and cooperate with relevant agencies. At the same time, most of DoD's metrics do not obviously draw on or tie into metrics developed by other agencies for similar or related purposes and related programs.
- The DoD Metrics Report deliberately does not consider future missions or changes in objectives, and for some programs it does not take into consideration or explicitly discuss planned and unplanned change over time. Projects have different phases that make different kinds of progress toward the program's impact and effectiveness. Also, circumstances change, including changes in the threats, the political environment, the program's priorities, and the funding available. A project's performance itself feeds back into the management and decision-making process and can lead to change. It will be difficult to address change and sustainability without considering and incorporating these factors explicitly.

The practical consequences of some of these shortcomings, such as not articulating the connection to threat or risk, might not be large for projects in progress under longstanding agreements. But they are important for good management, new projects, and especially for new partnerships.

IMPROVING METRICS

The committee has several recommendations to DoD on how it can improve its metrics for the CTR programs and reports on CTR. The committee does not recommend specific alternative metrics (DoD has to develop its own metrics with its partners), but does recommend what it considers a more effective approach for DoD to take to developing metrics. In the report, the committee provides an example of its recommended approach applied to the CBEP.

Objectives and Partnership:

1. **For each program in the DoD Metrics Report, DoD should include a concise statement of its objectives and of how the program is intended to reduce threat or risk.**
2. **Objectives for projects and the overall CTR Program in a partner country are developed jointly between the United States and the partner country. An agreed set of metrics should also be built into projects from the outset. They may change, but the parties responsible for the projects should know at any given time the metrics that will be used to measure impact and effectiveness.**

The CTR Program was established by Congress with clear authorities and each activity must begin with a clear statement of the United States' authorized overall objectives. DoD then needs to work with partner nations to define mutual objectives for their joint efforts. To measure impacts and effectiveness, metrics must include outputs (e.g., changes in apprehension rates at borders) not just input metrics (e.g., training materials provided).[6] Where possible, DoD should develop its metrics from outputs linked to the capacities that the programs are trying to build, and incorporate into the DoD-partner agreement provisions for metrics and the means to carry out those measurements.

As DoD takes CTR to new countries, it has opportunities to utilize lessons from 20 years of experience with cooperative threat reduction and develop metrics in a logical way, integrating them in the programs and in the projects from the beginning. The committee summarizes a logical order to developing metrics as follows.

1. State clearly the objectives of the overall U.S. CTR Program, including linkage to threat or risk.
2. Work with the partner country to define objectives for joint activities in their country. U.S. goals and partner-country goals do not need to be congruent (match exactly), but they must be compatible and should be explicitly stated.
3. Identify the partner capacity needed to meet the U.S. CTR Program objectives.[7]
4. Work with the partner country to define capacity development objectives for each capability. These may be prioritized based on their anticipated impact and the resources required to achieve good results (together yielding "bang for the buck").
5. Define metrics with the partner country based on capacity objectives. Agree on baselines, data milestones, and measures of success. Identify the source of data collected for each metric and who will provide and maintain the data. Ideally, an entity independent of the decision makers and implementers would collect and report on the performance. Different metrics may be appropriate for different stages.
6. Prioritize the metrics based on their importance to achieving the program objectives and the increase in capacity required.
7. Build metrics, including exercises if appropriate, into the implementation.
8. Provide metrics data in addition to time and costs expended for each project. Evaluate results independently (United States only) and together with partner country.
9. Feed evaluation back into the U.S. and partner country decision-making process.

Priorities

3. The committee judges that using a consistent framework to prioritize and refine metrics within a program would help DoD and other CTR decision

---

[6] It may not be possible to directly measure the higher-level outcomes from some CTR program's or project's performance, even with the best metrics. For example, DoD may never know how many illegal shipments were *not* interdicted at a border crossing assisted by a CTR program, or how many patients sick with an illness of interest did not go to the hospital. Indeed, it can be difficult to interpret the meaning of a change in a metric. Recognizing this fact at the outset will help to avoid wasted time and effort. These challenges are discussed in greater depth in Chapter 3 (see, for example, the section on Metrics for Weapons of Mass Destruction Proliferation Prevention Program).

[7] This mapping may be many to one, i.e., partner capacities may support one or more objectives.

makers. Using such a framework, DoD can identify the highest priority metrics, ensuring that the metrics are useable and useful, and allow decision makers to feed results back into the overall CTR objectives and budgetary process. Any of several decision-making or priority frameworks would work, including the decision analysis technique of swing-weight analysis and/or the DoD capabilities based planning process.

DoD developed each set of metrics in the DoD Metrics Report *ab initio*, using a different approach for each program, i.e. chemical, biological, borders, and nuclear. Although each CTR program is distinct and possibly unique, it is more difficult to be comprehensive, consistent, and focused without a consistent framework. An important component of measuring what matters is prioritizing among goals and within sets of metrics. Not all goals are equally important and the act of prioritization will help avoid double counting and other similar pitfalls. DoD should consider using a consistent framework to prioritize and refine metrics within a program. There are many decision and prioritization frameworks. In this report, the committee highlights the decision analysis technique called swing-weight analysis,[8] and the framework used widely in DoD called capabilities based planning. Using a consistent framework does not mean that the metrics for each CTR program or program element should be the same, but that a common framework should be used for defining program objectives, partner capacities, capacity objectives, and metrics.

Working with Other Agencies

4. **DoD plans to leverage other U.S. Government agencies' experience, capabilities, and assets as CTR expands to new countries and as it continues existing programs. DoD also needs to communicate, coordinate, and cooperate with relevant agencies.**

DoD is not the only agency engaged in capacity-building programs (such as those for deterrence, detection, and defense against WMD). U.S. Customs and Border Protection operates a program with equipment, training, and services similar to the CTR Proliferation Prevention Program along many thousands of kilometers of border and has developed metrics for its mission and operations. The U.S. Department of Agriculture's Animal and Plant Health Inspection Service is a leader in an international surveillance network that has many parallels to the global network DoD leadership envisions for reducing biothreats. The United States Agency for International Development operates capacity building programs with partners across the world to foster democratic institutions and has had to develop metrics for these hard-to-measure efforts. The DoD Defense Security Cooperation program also shares important similarities with the DoD CTR Program and may in some cases serve as a model.

DoD's CTR metrics do not address the full scope of the threat in each of the WMD areas. Instead, the metrics address only the scope of the funded projects. While the project metrics are useful in assessing the annual project status, they do not help Congress and senior leaders in

[8] A swing weight matrix defines the importance and range of variation for a set of metrics. The idea of the swing weight matrix is straightforward: a metric that is very important to the decision should be weighted higher than a metric that is less important. A metric that differentiates between alternatives is weighted more than a metric that does not differentiate between alternatives.

DoD and other parts of the government to understand the full scope of the potential for cooperative threat reduction which could help identify the need for and scope of future projects.

Not only can DoD CTR learn from these other agencies/programs, but DoD will be working with these other agencies in a "whole of government" effort. These and others might already have mechanisms in place for measuring impact and effectiveness that could be useful to DoD CTR.

Time and Change

5. **DoD's metrics and planning process should factor in more explicitly both planned and unplanned change over time. During the phases of active DoD involvement in a CTR project and afterward during sustainment, which is its own stage requiring resources (budgets, equipment, and trained people), clearer planning for how changes and metrics results will feed into decision making will make the metrics more credible and useful for both DoD and the partner country.**

The purpose of metrics is to provide information to inform CTR Program decision makers. Effectiveness and impact metrics should be tracked over time, and impact and effectiveness will be compared to the time and resources expended. In the early stages, even when operating according to plans, projects are unlikely to have measurable impact, but that does not mean that they are not on track or will not have an impact. Different project stages require different metrics as measures of progress. CTR Program managers and other U.S. Government decision makers need this information to ensure that project resources are achieving the program objectives. DoD and the partner country can also use this information to signal the need to develop and implement corrective action plans so resources can be better utilized or reallocated, if progress is not satisfactory. Explicitly factoring in change will help DoD with clarity and completeness of the metrics. When objectives change due to changing circumstances, managers may need to change the metrics as the original metrics may no longer be relevant. A phased, adaptive approach will enable DoD to use appropriate metrics at each stage of a project so that the right kind of impact and effectiveness can be shown even at the beginning of a project. This is especially true for sustainment.

Independent Evaluation

6. **Capacity building programs need independent evaluation of how the capabilities being built perform in action. This can be accomplished by several means, ranging from expert observations of routine operations to comprehensive exercises that test the full scope of capabilities. The level of effort can be tailored to the scope of the program, its resources, and its relative importance. DoD and its partners should build such independent evaluation into each project. The Defense Security Cooperation Program might be a good model for how to proceed.**

Independent evaluation establishes a degree of credibility that is hard to achieve by other means. Especially for capacity building programs, some kind of independent evaluation is essential. Exercises are a good way to measure effectiveness and sustainment. The kind of evaluation employed needs to be tailored to the scope of the program, its resources, and its relative importance. The evaluation might take the form of periodic expert observations of the project operations or the partner country's capacities. It might be an impromptu test of a randomly selected part of the system (e.g., a border protection system). Or it might be an exercise of the system, such as was performed for CBEP in Georgia as part of the Initial Operational Capability assessment. Ideally, the exercises would be designed by an entity independent from the groups being tested.

Measuring progress in building capacity and effectiveness of programs to prevent low-frequency, high-consequence events is difficult. Assessment against standards and guidelines (e.g., is the partner's action plan for interdicted nuclear material consistent with the International Atomic Energy Agency model action plan?) is one important component, but DoD and Congress care more about likely performance when the event occurs (i.e., how effectively can the partner implement the action plan?). Exercises can help measure both capability and performance in such programs. While exercises might be structured differently in different countries and for different projects, they should be built into the implementation and evaluation components of capacity-building projects and programs from the beginning, starting with a baseline evaluation and proceeding with midcourse and final or sustainment evaluations.

OTHER MAJOR ISSUES FOR CTR IN THE FUTURE

Some Congressional authorizers and appropriators have questioned whether CBEP should in fact be a part of DoD's CTR mission. It is beyond the scope of this study to comment on what is inside or outside the scope of the CTR mission. Fundamentally, whether DoD should be the agency to carry out the CBEP mission is only secondarily a metrics question. The primary question is whether the U.S. Government wishes to prioritize the work done under CBEP and what mix of government agencies is best equipped to carry it out. The increases in budget and scope to date, as well as the National Strategy for Countering Biological Threats, indicate that CBEP is a growing priority to the Administration and DoD's involvement reflects a conscious choice to use DoD to carry out this mission because of its experience with biodefense research and its experience working on threat reduction programs. Critics may dispute these decisions, and can legitimately point out that difficulties in developing reliable direct metrics for the program's impact and effectiveness increase the programmatic risks of the program, but mixing these issues with questions about the metrics themselves confuses matters and makes it more difficult to make progress in the program and in the debate about the program.

Finally, defining and measuring completion—how do we know when we are done?—and sustainability—will the improvements take hold and will the partner nation support and sustain the programs when U.S. funding stops?—are critically important for CTR programs, particularly capacity-building programs. What completion and sustainability mean and how they should be implemented and measured for a given program should be part of the formulation of objectives.

There is a mismatch in the vision of sustainability and measuring completion of the program among different CTR decision makers in the Administration and on Capitol Hill. One vision might be called a project view, in which DoD partners with a country, engages in a set of concrete activities with a well-defined beginning and end, and then DoD exits and monitors sustainment after project completion. The other main vision might be called a relationship view, in which DoD partners with a country, works with the partner to build a joint or multilateral network that is exercised regularly to maintain an on-going relationship with no defined end date. These visions appear mutually exclusive, but there are different phases to capacity building programs: the initial phase may involve intensive efforts and capital expenditures. There should be schedules and milestones for completion of this phase. The long-term relationship that follows may be open ended, but it also should require far less funding, which should allay some concerns about programs with no exit strategy.

1

Introduction

In the National Defense Authorization Act for Fiscal Year 2010 (P.L. 111-84), Congress directed the Secretary of Defense (DoD) to develop and implement metrics to "measure the impact and effectiveness of activities of the Cooperative Threat Reduction Program," (CTR) and directed DoD to request the National Academy of Sciences (NAS) to conduct "an assessment to review the metrics developed and implemented under subsection (a) and identify possible additional or alternative metrics, if necessary." (See Appendix A.) The Secretary completed a report describing DoD's metrics for the CTR Program (DoD, 2010; here called the DoD Metrics Report) in September 2010.[9]

As implied in the language cited above, a metric is an evidence-based tool that measures impact and effectiveness or performance of a program or a project toward its objectives. Metrics alone cannot ensure that the best options have been identified or are being implemented, but they can be helpful to program managers and overseers as they work to ensure that efforts are focused in productive directions. Impact and effectiveness in some projects is easily measured quantitatively, particularly where the objectives yield tangible products, such as constructing secure storage facilities, consolidating materials into a secure facility, or destroying a stock of a chemical agent. The impact and effectiveness of projects that have less tangible objectives (e.g., strengthening both a nation's ability to detect and respond to disease and the relationships between U.S. experts and those in the partner nation) can be more difficult to measure. Indeed, what to measure is an important but difficult question for such projects, and this was a central motivation for Congress's mandate for DoD to develop and NAS to review metrics for the CTR Program, which are described in the next section.

The committee's goal in this report is to assess the metrics DoD developed, and to provide recommendations on how to improve them or develop more effective alternative metrics. No metric will be perfect. Metrics are tools for evaluation, not substitutes for evaluation. Even in the missions with easy-to-understand, tangible actions, the obvious metrics (e.g., facilities completed, stocks of nuclear material consolidated) may not by themselves reflect the effectiveness that the United States cares about most (e.g., are the stocks secure?). With this limitation in mind, the committee provides background on the CTR Program, other studies of CTR, and the scope of this report before assessing the DoD Metrics Report.

The committee wrote this report with two main audiences in mind: Those who are mostly concerned with the overall assessment and advice, and those readers directly involved in the CTR Program, who need the details of the DoD report assessment and of how to implement the approach that the committee recommends. For the first category of reader, the committee has included a summary of the DoD CTR metrics in Table 2-1. For readers who desire more detail on the DoD CTR Metrics Report, the entire report can be found in Appendix B.

---

[9] See Appendix B for the DoD Metrics Report; Figure 1 of the Report shows the CTR Program within the DoD organizational structure.

BACKGROUND

The CTR Program, created in 1991, was based on an enlightened approach to national security conceived at a time of rapidly changing events surrounding the end of the Cold War: The United States and the Soviet Union (and its successor states) had a shared security interest in addressing the threat of nuclear weapons proliferation from the Soviet nuclear weapons complex at a time when the former Soviet Union (FSU) was experiencing profound political and economic crises. Because of the security risks precipitated by the unstable and evolving situation, the United States provided assistance to Russia and other states of the FSU to meet treaty obligations by dismantling weapon delivery systems and to address the challenges of improving protection, control, and accounting for nuclear weapons and materials; to work to prevent smuggling of nuclear weapons and their components; and to prevent transfer of actual weapons, components, and weapons-related knowledge to other countries or terrorist groups. How did the United States reach the point where it would provide such assistance to a set of new states that emerged from its former Cold War rival?

In 1990, the Soviet Union possessed tens of thousands of nuclear weapons, more than all of the other countries on Earth combined. The Soviet Union was also beginning to break apart. Several of its constituent republics declared independence over a 21-month period, a group of hard-line Communist Party leaders staged a failed coup attempt in August 1991, and in December 1991, the Soviet Union officially ceased to exist. Throughout the process of the Union's dissolution, nuclear security experts in the United States became increasingly concerned about the control of Soviet nuclear weapons in Russia and especially in other, newly independent states.

Recognizing the danger of leaving nuclear weapons on the territories of newly devolved states facing profound economic and political difficulties, Soviet President Mikhail Gorbachev asked U.S. President George H.W. Bush for assistance in dismantling some Soviet nuclear weapons and began removing Soviet weapons from military bases outside of Russia. When three heads of the former Soviet republics signed the Minsk Agreement on December 8, 1991, followed by eight others on December 21, formally dissolving the Union of Soviet Socialist Republics and forming the Commonwealth of Independent States, approximately 4,000 nuclear weapons remained in Belarus, Kazakhstan, and Ukraine. These agreements and the subsequent Lisbon Protocol signed with the United States on May 23, 1992, helped to ensure strategic stability by defining roles and future responsibilities under existing arms control treaties and establishing commitments for consolidation of Soviet nuclear weapons in Russia. Further steps were also needed to implement these commitments and to deal with parts of the former Soviet nuclear arsenal itself.

On December 12, 1991, President Bush signed into law The Soviet Nuclear Threat Reduction Act of 1991 (P.L. 102-228), commonly known as the "Nunn-Lugar" legislation, so named because Senators Sam Nunn and Richard Lugar were the sponsors of the bill. This legislation articulated two objectives for the CTR Program: "A) to facilitate on a priority basis the transportation, storage, safeguarding, and destruction of nuclear and other weapons in the Soviet Union, its republics, and any successor states; and B) to assist in the prevention of weapons proliferation." During the 1990s, the CTR Program expanded to address threats from the discontinued clandestine biological weapons programs and the stalled chemical weapons destruction program of the Soviet Union. These tasks have been substantial. The production scale of these Soviet programs was enormous: devastating chemical agents were stockpiled in thousands of tons; biological agents were produced on a massive scale. The CTR Program consolidated declared inventories of these agents, increased security at storage sites, and helped destroy and dismantle declared production facilities.

How has progress on cooperative threat reduction been measured? The "Nunn-Lugar scorecard," was developed to illustrate the impact and effectiveness using simple, direct measures.[10] It tracks the numbers of warheads and delivery systems dismantled and destroyed. As of January 2011, the scorecard reads as follows:

TABLE 1-1 Nunn-Lugar Scorecard, January 2011

7,599 strategic nuclear warheads deactivated
791 intercontinental ballistic missiles (ICBMs) destroyed
498 ICBM silos eliminated
180 ICBM mobile launchers destroyed
659 submarine launched ballistic missiles (SLBMs) eliminated
492 SLBM launchers eliminated
32 nuclear submarines capable of launching ballistic missiles destroyed
155 bombers eliminated
906 nuclear air-to-surface missiles destroyed
194 nuclear test tunnels eliminated
503 nuclear weapons transport train shipments secured
upgraded security at 24 nuclear weapons storage sites
built and equipped 20 biological monitoring stations
neutralized 1680.4 metric tons of Russian and Albanian chemical weapons agent

While this table is succinct and readable, text supporting the scorecard captures a greater accomplishment of the CTR Program: "Perhaps most importantly, Ukraine, Kazakhstan, and Belarus are nuclear weapons free as a result of cooperative efforts under the Nunn-Lugar Program. Those countries were the third, fourth, and eighth largest nuclear weapons powers in

[10] See http://lugar.senate.gov/nunnlugar/scorecard.html; accessed January 11, 2012.

the world."[11] Since 1991, much has been done; both accomplishments that are easy to present with a number or score and ones that are not have been achieved.

DoD Organization for CTR

Figure 1-1 illustrates how DoD has organized responsibilities for the CTR Program within the department, giving responsibility for overall policy decision making (what countries to partner with and what high-level objectives to pursue) to the Office of the Under Secretary of Defense for Policy, decisions about how to pursue the objectives to the Deputy Assistant to the Secretary of Defense for Treaties and Threat Reduction, and implementation responsibility to the Defense Threat Reduction Agency (DTRA), both in the Office of the Under Secretary of Defense for Acquisition, Technology, and Logistics. Different parts of the DoD Metrics Report that the committee reviewed may have been developed by different parts of DoD, but the committee has deliberately disregarded this. Congress tasked the Secretary of Defense to develop and implement metrics for the CTR Program, so the report is a DoD report, and the committee directs its comments and advice to DoD.

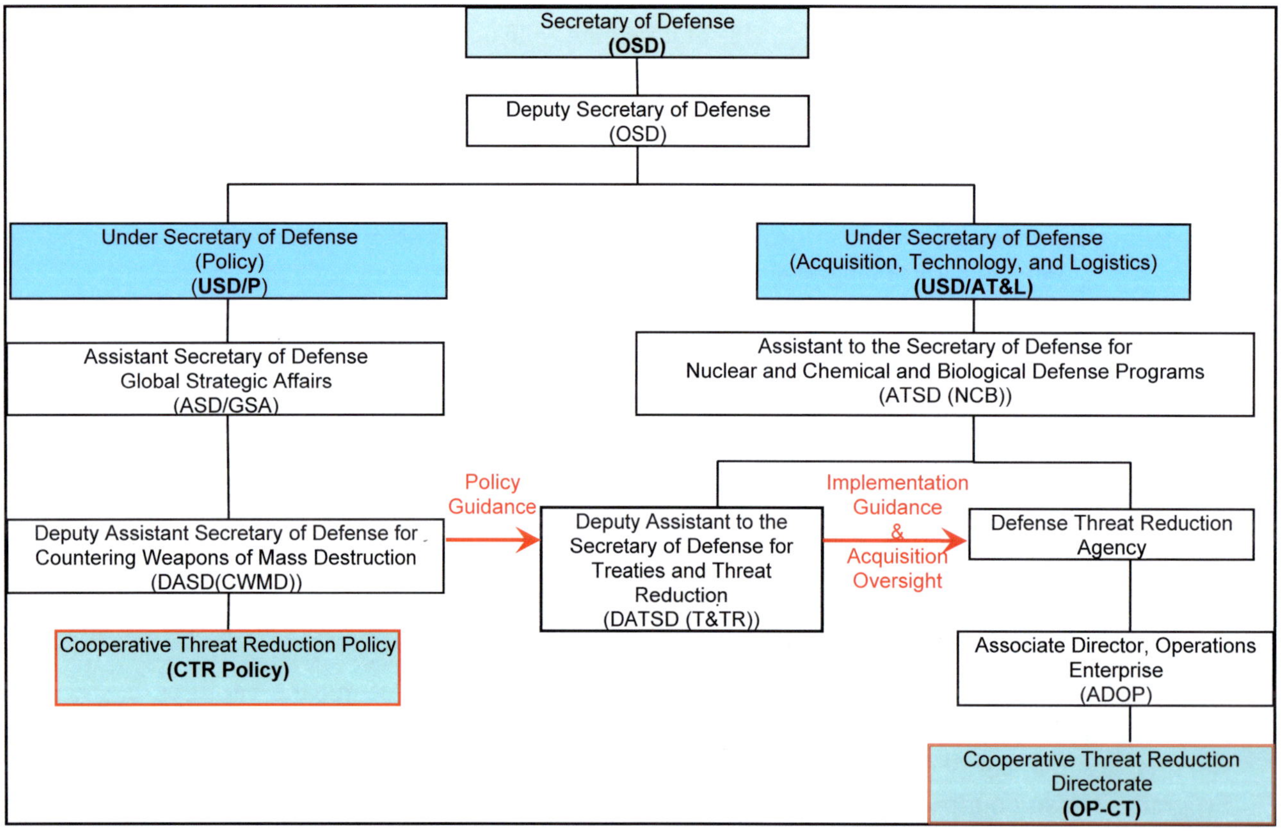

FIGURE 1-1 The CTR Program within the DoD organizational structure (DoD 2010).

[11] See http://lugar.senate.gov/nunnlugar/; accessed January 11, 2012.

PREVIOUS CTR REPORTS

Several studies have examined the CTR programs and given advice on how to make them more effective. Some of those studies are described in Appendix C. This section describes recent reports and developments.

The original legal agreement between Russia and the United States that provided the basis for the CTR Program is due to expire in June 2013, and there have been many efforts to define what, if anything, should follow that agreement. Moreover, given that the nature of security threats is different today than in the 1990s, some have considered how best to address these 21st century challenges through an expanded CTR Program. In the National Defense Authorization Act of 2008 (P.L. 110-181), Congress directed DoD to ask the NAS to recommend ways to strengthen and expand the DoD CTR Program. The report from that study concluded that

> Expanding the nation's cooperative threat reduction programs beyond the former Soviet Union, as proposed by Congress, would enhance U.S. national security and global stability. ... The committee recommends that the DoD CTR Program should be expanded geographically, updated in form and function....Forging broad new partnerships to implement sustainable programs that employ hard and soft capabilities and are tailored to specific countries or regions will energize and strengthen global security efforts and result in tangible and intangible benefits to national security. It is essential to develop meaningful program metrics that highlight program impact, acknowledge the value to national security of intangible program results, incorporate partner metrics into the overall evaluation of programs, and link metrics to program selection criteria. (NRC 2010)

In this sense, an expanded CTR Program is envisioned to expand not only beyond its core weapons of mass destruction (WMD) portfolio, and beyond the geographic regions in which it has thus far focused, but also beyond those measures of programmatic impact that have been utilized in the first decades of the Program.

Concurrent with that study, the White House developed its own strategy for countering biological threats (NSC 2009), which the federal agencies have used to guide their efforts (see, e.g., Weber, 2010).

Perceptions in Congress of the evolution of the Program and the challenges faced in the Program led to a congressional mandate for this study. The desire for new metrics for CTR is motivated by at least two concerns about the Program's growing work on capacity-building efforts aimed at improving partner nations' ability to deter, detect, and respond to emerging threats, particularly in the Cooperative Biological Engagement Program (CBEP). First, some national security constituencies question whether CBEP, which resides in a gray mission space that sometimes overlaps with public health, serves their national security goals and whether it should be carried out by DoD. These skeptics have seen the Biological Threat Reduction Program shift from efforts to redirect weapons scientists and Biopreparat (the Soviet biowarfare agency) to legitimate commercial ventures, to efforts to engage nonweapons bioscientists in Central Asia, Africa, and even India,[12] and ask DoD to define the *security* value of today's capacity building health and biology related programs today.[13] Second, impact and effectiveness

---

[12] See, for example, Weber (2010) and Royal (2011).

[13] The committee heard skeptics of CBEP say that unless the partner country's scientists or their programs are weapons programs, the engagement is not a defense mission and CBEP does not belong in DoD. DoD argues that biological threats have changed greatly in the last 20 years. Proliferation of ability to create dangerous biological agents has already happened globally, threats are asymmetric, and only a small amount of material is needed to

success are difficult to measure in such efforts, and appropriators and program managers alike want to ensure that the implementers have accurate, timely feedback on progress to guide the Program, and that funds spent on the Program are expended effectively.

BRIEF BACKGROUND ON DOD CTR METRICS REPORT

DoD issued a report on metrics for the CTR programs titled "Report on Metrics for the Cooperative Threat Reduction Program, September 2010." (See Appendix B).[14] The DoD Metrics Report consists of a brief introduction followed by descriptions of the metrics for all but one of the CTR programs: CBEP Program, the Chemical Weapons Elimination Program, the Nuclear Weapons Safety and Security Program, and the WMD Proliferation Prevention Program. The Strategic Offensive Arms Elimination Program was not included because that program uses the "Nunn-Lugar Scorecard" metrics, and will continue to do so until that mission is completed.

SCOPE OF THIS ACADEMY REPORT

This report provides the study committee's assessment of the metrics in the DoD Metrics Report, including the feasibility of using the metrics (can these measurements be made?), their alignment with the objectives of the program (do these metrics reveal progress toward the stated objectives of the program?), and their usefulness to people with responsibility for implementing, directing, overseeing, or otherwise supporting the program. To the extent that there are shortcomings in the proposed DoD metrics, the committee tries to provide useful suggestions for how to improve the metrics and the use of metrics overall. The committee does not recommend specific alternative metrics, but does recommend what it considers a more effective approach for DoD to use in developing metrics. DoD has to develop its own metrics with its partners. In the report, the committee provides an example of its recommended approach applied to CBEP.

---

create a biological weapon that could cause billions of dollars of damage to our economy (one expert said that amount could even be "pocket-sized"). With the capability already spread widely, intent is more important now, and to know and affect intent the program must communicate with capable and knowledgeable individuals with whom we have built some degree of understanding and, in some cases, even trust.

[14] The DoD Metrics Report is refreshingly brief, but unusual in that it has no indication of who wrote the report, and has no trappings of an official document (e.g., the agency name or seal, document number).

2

Committee Assessment of the Department of Defense Cooperative Threat Reduction Metrics Report

The first step in developing metrics is to establish what questions the metrics are meant to answer, and how the metrics will be used. Throughout this study, in conversations with the Department of Defense (DoD), other agencies, staff on Capitol Hill, and in the Executive Office of the President, and with officials in partner countries, the committee asked how the metrics will be used. The answers ranged based on the respondent, from those who sought measures of progress toward agreed goals, to cost effectiveness, to others who sought measures of how well the efforts align with the national defense mission. Given that different audiences want metrics for different purposes, it makes sense to begin by describing the committee's perspective on the purposes of metrics.

Metrics are tools for evaluating the impact and effectiveness of programs and projects against strategic goals *and* for management within the program. The former are program metrics. The latter are project metrics. Different metrics may be appropriate for different stages of program and project development, but they all need to be tied together by the strategic objectives of the program as they fit with broader DoD Cooperative Threat Reduction objectives.

The committee evaluated the DoD CTR metrics described in the (CTR) DoD Metrics Report based on whether the metrics provide decision makers with the essential information to manage the effectiveness and impact of CTR programs. The committee drew on several frameworks for program planning, performance, and evaluation, but especially on capability based planning, to identify general steps needed for meaningful evaluation.[15] Other frameworks are also valid and will either explicitly or implicitly include these elements. DoD must

- state the objectives of the program and the project (i.e., the goals of the actual activities)
- identify the capabilities it is trying to develop or maintain
- define objectives for each capability and link those capabilities to metrics
- ensure that the metrics reflect program effectiveness and impact
- plan for and measure sustainment (see footnote 3 in the Summary)

It is generally good practice for the program to establish minimum performance levels and aspirational goals for each metric (see footnote 4 in the Summary). In Table 2-1, the committee

[15] A much more elaborate metrics framework is emerging from analysts focused on the international development community. This highly structured approach treats the portfolio of projects as an experimental trial, with control projects or communities and strict adherence to a protocol established at the outset of the program or project. The committee chose a program evaluation framework more like the program planning, performance, and evaluation approaches already used in DoD because the latter are more familiar to DoD and, in the committee's view, better suited to the CTR Program. The CTR Program engages in too few projects for meaningful control groups, and the political conditions, physical and social circumstances, budgets, and objectives for the program change, all of which make a trial-based approach incompatible with CTR today.

TABLE 2-1 Summary of metrics from the DoD Metrics Report

CTR Program	What metrics are intended to measure	Metrics
Cooperative Biological Engagement Program (CBEP)	1. Secure and consolidate collections of especially dangerous pathogens (EDP) and their associated research at a minimum number of secure health and agricultural laboratories or related facilities. 2. Enhance partner country/region's capability to prevent the sale, theft, diversion, or accidental release of biological weapons (BW)-related materials, technology, and expertise by improving biological safety and security (BS&S) standards and procedures. 3. Enhance partner country/region's capability to detect, diagnose, and report endemic and epidemic, man-made or natural EDPs, bio-terror attacks, and potential pandemics. 4. Ensure the developed capabilities are designed to be sustainable within each partner country/region's current operating budget. 5. Facilitate the engagement of partner country's/regional scientific and technical personnel in research areas of interest to both the partner country/region and the United States. 6. Eliminate any BW-related infrastructure and technologies encountered in a partner/country region.	*Examples of measures of effectiveness*[a] Partner country EDP collections and associated research are consolidated into a minimum number of locations. Partner country EDPs and associated research are secured in a manner consistent with standards. Partner country has BS&S laws and regulations governing work with EDPs. Partner country disease detection and diagnosis capability meets U.S. and/or international guidelines for biosafety. Partner country has preparedness and response plans. Partner country disease surveillance system is capable of detecting and reporting suspect EDP cases to those responsible for human and animal health.
Chemical Weapons Elimination Program (CWE)	Destruction of agent through the Russian Federation's safe, efficient operation of the destruction facility	Quantity of CW agent destroyed Number of munitions destroyed Project Metrics Scheduled facility downtime Unscheduled facility downtime Facility achieved availability
Nuclear Weapons Safety and Security (NWSS) supports Nuclear Weapons Transport Security Program	1. Performance/capability assurance 2. Configuration management 3. Procedures and process 4. Training 5. Organization and personnel 6. Life-cycle management 7. Maintenance 8. Logistics	*Examples of metrics* Number of nuclear weapons storage sites upgraded Number of personnel trained Establishment of regional technical centers Note: Some metrics still under development
Weapons of Mass Destruction Proliferation Prevention Program (WMD-PPP)	1. Enhance partner country capability to perform effective risk management 2. Enhance partner country capability to perform border security command, control, communications and computers 3. Enhance partner country capability to perform border security surveillance 4. Enhance partner country capability to perform WMD detection 5. Enhance partner country capability to perform border security interdiction 6. Assist partner country with the development of a sustainment budget for all systems delivered under this program 7. Enhance partner country capability to support and maintain delivered equipment and/or systems 8. Enhance partner country capability to sustain delivered training 9. Enhance a partner country capability to capture and disseminate information including that concerning WMD incidents 10. Increase the awareness of partner countries as to their critical role in WMD non-proliferation (partner country buy-in) 11. Develop interagency, bilateral, regional, and multilateral cooperation	*Examples of metrics* Miles of green (land) / blue (maritime) border with added security Number of ports of entry enhanced Number of border facilities (land and maritime) provided increased capability to detect WMD Number of personnel trained and equipped to perform border security interdiction Initial and refresher training system exists Number of personnel exchanges (training, professional development) Mean time to negotiate agreements Number of regional relationships facilitated

[a] The terms "goal," "objective," "indicator," "attribute," and "metric" are used inconsistently among the different program descriptions in the DoD Metrics Report. CBEP uses the term "measure of effectiveness" to describe what is being measured and the term "indicator" to denote how it is being measured (i.e., what is being counted).

attempts to summarize the disparate metrics from the DoD Metrics Report and Table 2-2 summarizes the committee's assessments of DoD's CTR Metrics Report with respect to the elements described above. Because of differences among the programs, inconsistencies in their terminologies, and the complexity of their metrics, the program metrics and indeed the assessments are difficult to represent in a uniform tabular structure, so the committee commends readers seeking deeper familiarity with the metrics to the DoD Metrics Report, which is reproduced in full in Appendix B. The committee did not address the Strategic Offensive Arms Elimination Program, which is not discussed in the DoD Metrics Report. DoD indicated that it plans to use long-standing "Nunn-Lugar Scorecard" metrics for that program.

METRICS AND OBJECTIVES

In its introduction, the DoD Metrics Report states that "We can measure the amount of equipment provided and the number of training events conducted or scientists engaged; however, we need better measures to show that these efforts actually result in changed practices or additional effectiveness." The committee agrees and notes that the example given conflates inputs and project metrics (those that measure progress on the means to meet objectives) and outputs and programmatic or strategic metrics (those that measure progress relative to/toward the objectives themselves).

Effectiveness and efficiency are important criteria for the evaluation of government programs. A government program is effective if it achieves the objectives set forth by the partners engaged in the program.[16] A government program is efficient if it minimizes the resources required to achieve its goals. A common approach is to manage and assess programs based on their cost-effectiveness, which is an efficiency measure. Although measures of effectiveness and impact are the most challenging and should continue to receive the most attention, measures of efficiency and cost-effectiveness are also necessary and should be included in future iterations of the CTR Program metrics.[17]

OVERALL ASSESSMENT

The DoD Metrics Report contains reasonable metrics for the CTR programs aimed at consolidation and elimination of weapons and weapons materials, and contains a reasonable starting point for developing metrics for the newer capacity-building programs.

---

[16] It may not be possible to directly measure the higher-level outcomes from some CTR programs or projects, even with the best metrics. For example, DoD may never know how many illegal shipments were *not* interdicted at a border crossing where a CTR program provides assistance, or how many patients sick with an illness of interest did *not* go to the hospital. Recognizing this fact at the outset will help to avoid wasted time and effort and avoid false expectations.

[17] While cost-effectiveness is important, particularly as a part of prioritization, it is essential that efficiency (which is relatively easy to measure) not drive out effectiveness (which is more difficult to measure). In other words, programs that are extremely cost-efficient may be pursuing the wrong result, or be solving the wrong problem. Balancing cost-effectiveness against other measures is a necessary part of developing prioritized metrics.

Individual Program Objectives

The DoD Metrics Report describes both CTR's highest-level objectives and DoD's difficulties in developing metrics, and it does so clearly and succinctly in the introductory section of its report. However, the individual program objectives (those of the CWE, NWSS, CBEP and WMD-PPP) and their connections to threats to U.S. national security are not well articulated or not addressed. Each program has objectives, and they may be described well in other DoD documents (see, e.g., Nacht, 2009), but the DoD Metrics Report itself needs to articulate the objectives in a way that makes apparent the connection between the actions taken under the Program and the higher level objectives. Not making those connections poses a problem for the Program because proper metrics indicate progress toward objectives, or more specifically, progress in developing the capabilities to fulfill the objectives. Those connections are the logical starting point for the presentation and for the actual development and analysis of metrics. Again, this is not to say that there is no connection, but that the DoD Metrics Report does not describe the connection.

For example, describing the connection would clarify how building the capacity to better track respiratory disease in East Africa reduces threats to U.S. national security or why Ukraine's interdiction of smuggled frozen chicken legs indicates a positive impact from a CTR program. A strong case can be made for each of these, but not without stating the Program objectives and connect them to program objectives.

For the programs that do not already have such objectives stated in the DoD Metrics Report, a concise statement of the objectives of each program and how the actions planned under the programs are intended to reduce threat or risk would provide a connection between the objectives and the metrics.

Partnership

CTR programs are intended to be and work best when they are carried out as partnerships (*cooperative*) with other countries. This includes joint development of the objectives with the partner countries. Most of the metrics in the DoD Metrics Report do not reflect such joint development. The report reads as if metrics are U.S. measurements of our partner's progress toward U.S. goals. This can create misunderstandings between the United States and our partners, which can undermine CTR efforts.

Prioritization

DoD did not use a consistent framework for developing and articulating objectives, priorities, and metrics, and DoD did not prioritize among its metrics. As a consequence, the DoD Metrics Report mixes project management measures with higher-level program performance metrics for some of the CTR programs, and weights equally metrics that are critically important and others are not.

Cooperation

DoD plans to leverage other United States Government agencies' experience, capabilities, and assets as CTR expands to new countries and as it continues existing programs.

DoD also needs to communicate, coordinate, and cooperate with relevant agencies. Examination of other U.S. Government documents suggests that CBEP is the most consistently planned and coordinated across the other agencies.[18] The committee notes that previous Academy reports and government agencies themselves have stated the importance of such coordination for program success, so a metric for it seems reasonable and valuable in theory. To the committee's knowledge, there is no metric to reflect interagency coordination or a "whole of government" approach to working with a partner country, rather than a piecemeal approach. The committee does not have a solution to this problem (see Chapter 3 for further discussion).

In addition to working with these other agencies, DoD CTR can learn something from them. They, too, engage in capacity building programs, work on similar missions, and measure performance. U.S. Customs and Border Protection conducts a mission similar to the PPP along many thousands of kilometers of border and has developed metrics for its mission and operations. The U.S. Department of Agriculture's Animal and Plant Health Inspection Service is a leader in an international surveillance network that has many parallels to the global network DoD leadership envisions for reducing biothreats. The U.S. Agency for International Development operates programs with partners across the world to foster democratic institutions. As a further example, a partner's compliance with the International Health Regulations would reflect interests and activities of DoD, Health and Human Services, and Department of State, and the extent of their coordination and cooperation in achieving that outcome is another metric.

Even other organizations within DoD could provide different models: The Defense Security Cooperation Agency builds security capacities with partner nations. DoD's CTR Program can learn from these other agencies successes and challenges (and even its own experience) if it incorporates a lessons-learned mechanism for the Program. Furthermore, the CTR Program will be working with these other agencies in a whole-of-government effort, and the others might already have mechanisms in place for measuring impact and effectiveness that would be useful to CTR.

Change

The DoD Metrics Report deliberately does not consider future missions or changes in objectives, but it is difficult to see how CTR metrics can be designed to respond to change if this is not discussed in DoD's report. Some factors are under the Program's control. Far more are outside of the Program's control. Every responsible business, military operation, and government program builds in resilience that is the ability to deal with exogenous change. The CTR programs are susceptible to external change ranging from budget reductions to shifts in the host country's political or economic environment.

The practical consequences of some of the shortcomings listed above, such as not articulating the connection to threat or risk, might not be large for projects in progress under longstanding agreements. But they are important for new projects and especially for new partnerships. A more detailed examination of the programs can be found in the next section.

[18] See Weber, A. 2010.

PROGRAM-BY-PROGRAM ASSESSMENT

The committee has summarized several important aspects of its assessment of the DoD Metrics Report in Table 2-2. The column headings represent a logical chain from objectives through metrics to sustainability and the minimum performance required, read from left to right.

Chemical Weapons Elimination (CWE)

The treatment of the CWE program in the DoD Metrics Report may be adequate for the current project under the program. The stated program goal looks very much like a project goal: "[T]o destroy nerve agent-filled munitions located in the Planovy [chemical weapons] storage facility in a safe, secure, and environmentally sound manner." The United States and the Group of Eight Global Partnership against the Spread of Weapons and Materials of Mass Destruction (G8 Global Partnership) supported the design and construction of the chemical weapons destruction facility at Shchuch'ye, Russia, which began operating in 2009. At Russia's request, the CWE program is providing technical support for the commissioning and operation of the facility. So the desired capability to support the program and project objectives is defined as adequate operation (by the Russians with U.S. technical support) of the Shchuch'ye destruction facility.

In its Metrics Report, DoD refers to the quantities of chemical weapons agent destroyed (neutralized and immobilized) as the program metrics. DoD's only project level metrics are scheduled facility downtime and unscheduled facility downtime.

These metrics align well with the "destroy nerve agent" aspect of the stated program objectives, but the metrics only implicitly address the "safe, secure, and environmentally sound manner" aspect of the objectives. Worker safety and environmental emissions are also important indicators of proper operation of the facility, but are not mentioned in the DoD Metrics Report. In a facility that has strong regulatory controls and/or strong internal management controls on safety and environmental performance, problems with safety or environmental releases affect the operation of the facility, i.e., the facility will shut down some or all of its operations to correct problems. If the facility does not have strong controls in place, safety and environmental performance might not affect continued operation of the facility, so tracking facility downtime might not factor in safety and environmental performance. DoD is better positioned than the committee to assess this aspect of the Shchuch'ye facility and can justify its decision either to include such metrics or to exclude them. Likewise, the neutralization process is supposed to convert 99.99 percent of the agent to slightly hazardous components and then immobilize them for disposal. DoD does not explain the acceptable range of destruction percentage (is 90 percent destruction acceptable?) and whether it is important to track that parameter.

Finally, the committee notes that the Planovy Chemical Weapons Depot near Shchuch'ye contains approximately one seventh of the Russian stockpile of chemical weapons that awaits destruction. If DoD hopes that the support provided at Shchuch'ye will not only enable the Russian Federation to operate Shchuch'ye independently, but also affect progress at other depots, then DoD should include a metric for that objective too.

TABLE 2-2. Program-by-program summary of objectives, capabilities, and metrics in the DoD Metrics Report. The committee's assessment (shaded) is followed by a summary of the committee's advice

	Objectives (program/ project) are specified?	**Desired capabilities identified?**	**Linkage of capabilities to metrics**	**Do metrics measure impacts and effectiveness?**	**Do metrics address sustainability?**	**Are minimum performance & aspirational goals specified?**
Chemical Weapons Elimination (CWE) Assessment	Not in DoD Report;[a] Some confusion between program and project objectives.[a]	Adequately operating facility Shchuch'ye	Yes	Yes – for the operation of Shchuch'ye	None	Yes
CWE Advice	N/A	Consider whether development at other facilities is needed.	This linkage needs to be established.	Worker safety & plant emissions may also be appropriate & important metrics.	This aspect could be developed.	N/A
Nuclear Weapons Safety & Security (NWSS) Assessment	Yes/Yes	Yes, for sustainment[b]	Yes	No, only measure input.	Yes	Yes[a]
NWSS Advice	Clearly link overall objectives to threat reduction.	N/A	This linkage should be made more explicit, particularly as the program expands to other countries.	Although there are constraints on measurement, improved metrics should have more focus on impact & effectiveness.	N/A	N/A
Cooperative Biological Engagement Program (CBEP) Assessment	Yes/Yes	Yes	Yes	Some, but not prioritized.	Yes	Yes
CBEP Advice	Need to be more clearly linked to threat reduction[a]	N/A	Would benefit from linking capabilities metrics with threat reduction.	Need to be prioritized, reduced in number, and direct metrics, where possible.	N/A	N/A
Weapons of Mass Destruction Proliferation Prevention Program (WMD-PPP) Assessment	Yes/Yes	Yes, but vague.	Program metrics not linked to capabilities. Project, yes	Program: no. Project: Some of them do, but not prioritized, not clearly linked to program level.	Yes	No
WMD-PPP Advice	Recognize links to other related/similar programs; address integrated assistance efforts across the USG and other aid providers.	Need to be clarified.	Once capabilities are better articulated, they need to be better linked to the program and project metrics.	More thought should be given to prioritization and impact and effectiveness.	Sustainability metrics also need to be prioritized and focused on capabilities.	The minimum performance and aspirational goals need to be specified.

[a] Not available in the DoD Metrics Report but available in other documents.

[b] NWSS has completed upgrades at agreed facilities, so the current program is focused on providing sustainment.

Cooperative Biological Engagement Program (CBEP)

For the biosecurity mission, there are two main thrusts:[19] (1) secure and consolidate especially dangerous pathogens, and any work with those pathogens, to a safe, secure facility or facilities; and (2) engage bioscience institutions[20] and authorities in the partner country to establish a culture of responsible practice, detection, and international reporting of emergent pathogens and transfers of the pathogens. In principle, the first thrust is relatively easy to measure, if the work with those pathogens is known to the governments. The second thrust, the aim of which is in part to develop relationships of trust and practices that are trustworthy, is more difficult for the United States to measure (see Measuring Trust, Confidence, and Goodwill in Chapter 3). The metrics DoD developed for CBEP are numerous and mostly indirect with respect to trust, but relatively concrete and measureable. Some additional steps are needed to make these metrics effective tools for evaluating impact and effectiveness and managing the programs.

The DoD metrics for CBEP have nearly all of the elements that the committee thinks are needed for development of useful metrics. The CBEP objectives and desired capabilities are clearly stated, although their connection to threat reduction is not stated in the DoD Metrics Report. The metrics are linked to the desired partner country capabilities, and some of them measure impacts and effectiveness, although others do not.[21] DoD also factors sustainability into its rating of the program's performance.

The greatest shortcoming of this otherwise rather complete metrics structure is its lack of prioritization. DoD lists 49 metrics for CBEP and in its first application of the metrics, DoD weighted all of them equally. It is difficult to draw meaningful conclusions about overall progress from 49 metrics that are all weighted as equally important. Some of the metrics are far more important to DoD and to threat reduction than are others (see, e.g., Footnote 7 in this chapter), and not all 49 are needed, even after duplicated metrics are eliminated. The committee discusses metrics for the CBEP more extensively in Chapter 3, where the committee illustrates how an improved method for developing, prioritizing, and using metrics should be applied to CTR programs.

---

[19] This is a summary of the four objectives DoD describes in its Metrics Report:

1. Secure and consolidate collections of especially dangerous pathogens (EDPs) and their associated research at a minimum number of secure health and agricultural laboratories or related facilities;
2. Enhance partner country/region's capability to prevent the sale, theft, diversion, or accidental release of biological weapons related materials, technology, and expertise by improving biological safety and security standards and procedures;
3. Enhance partner country/region's capability to detect, diagnose, and report endemic and epidemic, man-made or natural EDPs, bio-terror attacks, and potential pandemics; and
4. Ensure the developed capabilities are designed to be sustainable within each partner country/region's current operating budget.

[20] It seems reasonable to enhance this priority to include engaging bioscientists, not just bioscience institutions.

[21] For example, one of the measures of effectiveness concerns the partner country laws and regulations for biological safety and security, and their conformance to U.S. standards. While the U.S. implementers may believe that having such laws and regulations is better than not having them, the actual practices are more important than the details of the laws. The practices may be poor despite good laws and regulations or good despite poor laws and regulations. As a result, a partner that otherwise has shown significant improvements with respect to biological safety and security might get a poor rating based on a regulatory system that is different from that in the United States.

Nuclear Weapons Safety and Security (NWSS)

The NWSS section of the DoD Metrics Report gives candid assessments of the metrics it has developed (e.g., "*Depending on the level of cooperation, DoD may not be able to track this metric independently.*"), and overall gives the impression of a metrics effort in development. This reflects the current state of the NWSS program, but it also gives an incomplete picture of DoD's thinking about measuring the impact and effectiveness of this program.

NWSS has completed upgrades at all of the facilities covered by the agreement between the United States and the Russian Federation, and DoD plans to continue to use the number of nuclear weapons storage sites upgraded as a metric of the impact and effectiveness of the program in Russia. Given that the upgrades are done, the current program is focused on (1) providing sustainment in the form of training, and on (2) the Nuclear Weapons Transportation Security Program. The DoD Metrics Report also states that "with the potential global expansion of the NWSS program to other partner countries, Defense Threat Reduction Agency DTRA has been working to develop enhanced metrics that better reflect how our efforts are contributing to overall threat reduction."

The current NWSS program has severe constraints on measurements. The CTR program trains people who train the operators of the nuclear weapons storage sites, but the CTR training takes place at facilities away from the nuclear weapons storage sites, and CTR personnel do not have access to any of the sites. CTR supports supply of upgraded secure railcars for the transportation mission, but CTR personnel do not see them in operation. In both cases, the lack of access makes it difficult or impossible to measure the outcomes directly linked to the program's objectives.

Recognizing these constraints, DoD has identified the objectives of the program and the capabilities desired to achieve those objectives. The DoD Metrics Report describes what DoD would like to measure and what might actually be possible. For example, DoD will track the number of railcar shipments supported and the number of secure railcars provided. DoD can measure the latter itself; it cannot measure the former. DoD included some metrics in its report because they were negotiated among DoD, Department of Energy, and the partner, Russia's Ministry of Defense. Such mutually agreed metrics are important and notable, as discussed elsewhere in this report.

DoD presented additional information and ideas to the committee in November 2010.[22] These ideas include a systematic method for weighting or prioritizing among the lower level factors that contribute to the metrics. Without endorsing the specific details presented to the committee, the committee agrees that DoD would benefit from further development of ideas like those in the November presentation.

Weapons of Mass Destruction Proliferation Prevention Program (WMD-PPP)

The WMD-PPP focuses on land and maritime border security, also known as green and blue borders respectively, in partner nations including Azerbaijan and Ukraine. Three program metrics are defined:

- Miles of green (land) border provided sustainable security enhancements (in progress/complete)

[22] See Appendix E.

- Miles of blue (maritime) border provided sustainable security enhancements (in progress/complete)
- Number of ports of entry provided sustainable capability enhancements (in progress/complete)

In introducing these metrics, the DoD Metrics Report notes: "The ability to measure simply and objectively the impact that WMD-PPP assistance has had on threat reduction is challenging due to the nature of the program: we are providing a capability to our partners that gives them an ability to deter proliferation."

DoD articulates both program and project objectives for the WMD-PPP. The difficulty for WMD-PPP program managers and decision makers results from the vagueness of the capabilities desired to achieve the goals or objectives. DoD clearly found it difficult to translate the desired capabilities into metrics. Miles of green and blue border with additional security enhancements, and the number of ports of entry where capabilities have been enhanced are ill defined. Are all enhancements equivalent? Is every kilometer of border equivalent in terms of threat reduction?

Securing a border is complicated. The first approximation of progress—the number of kilometers of green border deemed secure—might not accurately represent risk reduction for several reasons. First, not all kilometers of border represent equal risks. Second, adversaries can adapt to exploit the weakest link. Third, what constitutes a secure kilometer of border? Areas where security enhancements have been provided might still have known smuggling routes that regularly defeat security enhancements may be improperly counted in success measures. These challenges are well known to DTRA, and in implementing WMD-PPP, DTRA has not treated all borders as if they were equally important. For instance, some stretches of border may be sufficiently impassable due to the natural topography of the area as to render any additional security enhancement unnecessary while others may include commonly used smuggling routes. Metrics should instead focus on the risk posed and the likelihood of evading detection. In Ukraine, both DTRA and the Ukrainian State Border Guard Service conducted a simple risk ranking and agreed jointly on the highest risk stretch of border. They focused their prototype and demonstration facility efforts on that stretch of border. Thus the implementation is better than the metrics would suggest.

Interdiction events involving materials useful for WMD are rare and experts are cautiously optimistic that this is because smuggling of WMD material is rare, not just difficult to observe. Border protection services do not now have, and endeavor to never have, enough WMD incidents to make tracking them a useful statistical tool, so they use statistics from other border violations as proxies to indicate the effectiveness of their border controls. The definition of positive outcome must be sufficiently broad to prevent unintentionally causing a reduction of inspections or alarms that might prevent the detection of an illicit item. For instance, if the risk management device is using rules based targeting efforts or non-intrusive inspection equipment, identifying a shipment that comes from a legitimately high-risk source or known bad actor might be a positive outcome, even if the resulting inspection did not uncover illicit material. Similarly, the physical inspection of a container that uncovers a significant quantity of lead or similarly dense material may be a positive outcome to imaging analysis of a container x-ray that shows a density anomaly. Noting those inspections as negative outcomes may result in the overly narrowing of risk factors, thereby increasing the risk of missing an actual smuggling event.

Progress in securing a border can be measured statistically (with some degree of accuracy, Chapter 3) if the country has sufficient control or awareness to even estimate the violations. Border protection services can track radiation alarms and smuggling incidents involving drugs, firearms, people, stolen goods, and goods avoiding tariffs.

It may be that partner countries do not need the ability to perform maritime interdictions themselves. Instead, one might measure the response time to conduct a maritime interdiction, understanding that another country's team or a regional/multi-national team may be best situated and equipped to respond. Measuring the time for interdiction might more accurately measure the objective. For instance, the coastline of Georgia may be such that it needs fewer boarding teams than required to respond adequately to the much larger coastline of Ukraine. Another option might be to take the numerator/denominator approach, measuring the current capability against an assessed need.

At a project level, the metrics in the DoD Metrics Report are better linked to the capabilities desired. Some of the project metrics DoD developed for the WMD-PPP measure impacts and effectiveness, such as alerts resulting in a positive outcome, and others do not, such as the number of maintenance personnel trained. There are also a number of other programs of the United States and other countries that provide border control and law enforcement assistance that affect border control and trafficking that may provide useful insights for DoD.

METRICS AND EVALUATION

Metrics are inputs to evaluation, not evaluations themselves or substitutes for evaluations. The committee cautions those who use program metrics to use them judiciously. Some people will always trust quantitative or numerical metrics over qualitative metrics in a belief that they are somehow more rigorous or objective. But numerical metrics are not necessarily objective and especially when taken by themselves can be misleading because different countries will have different baselines and different objectives. Likewise, it would be unfortunate if otherwise appropriate attention to metrics results in funds being taken out of activities that are useful and shifted into activities that are measurable.

The WMD-PPP illustrates some of the caveats that must be kept in mind with quantitative metrics. In addition to the shortcomings to the "miles of border secured" metric described above, using a more nuanced quantitative evaluation of border security based on interdiction statistics requires interpretation beyond the numbers. A country that has poor control of its borders or poor ability to interdict illegal trafficking through its border crossings (either because of inability to detect illegal trafficking or because of corruption) may have statistics tracking its progress in securing its borders, but those statistics will require interpretation. Does a rise in interdictions indicate higher risk (more trafficking) or lower risk (more traffickers getting caught)? See Chapter 3 for a more general description of what metrics cannot do.

3

Improvements to CTR Metrics

INTRODUCTION AND GENERAL DISCUSSION OF METRICS

In this chapter, the committee begins by discussing metrics in general, including terminology and a kind of taxonomy of metrics, what metrics cannot do, and some thoughts on measuring trust and confidence. In the recommendations section, the committee offers recommendations about objectives and partnership, working with and learning from other agencies, and prioritization. Following the first set of recommendations, the committee gives specific advice concerning border security (the Weapons of Mass Destruction Proliferation Prevention Program (WMD-PPP), and the Cooperative Biological Engagement Program (CBEP) program because the programs illustrate the points well.[23] Those sections are followed by recommendations on independent evaluation and factoring time and change into metrics. Chapter 3 closes with a set of important issues for Cooperative Threat Reduction (CTR) that are sometimes seen as metrics issues, but in fact are not.

On Metrics

Metrics are used in many private and public domains: strategic planning and assessment; business planning and assessment; manufacturing and service planning and assessment; policy analysis; campaign planning and assessment; product/service design and assessment; systems analysis; systems engineering, and engineering management. The term "metrics" is very broad and additional terms are useful to differentiate the full range of metrics that are needed for a large, important program like the CTR Program. A list of the most common terms used for metrics can be found in Box 3-1.

The terms in Box 3-1 provide a much richer understanding of the purposes and the types of metrics. Unfortunately, there is significant overlap in the terminology. Systems thinking can help to organize similar terms into categories that provide a reduced set of terms and allowed the committee to perform its assessment of the current CTR metrics and identify opportunities for improvement. Figure 3-1 shows six categories of metrics: input, process, environmental, output, value, and benefit cost metrics. The term on the top of each category is the term the committee uses in the report to assess the CTR metrics. The terms in parentheses are similar terms binned in the categories. Leading and lagging indicators are shown because they are an alternative way of grouping the six metric categories based on the timeliness of the metric for determining corrective action.

In addition to the purpose of the metric and the category of metrics, another important distinction is what the committee will call the "type" of metric or measure. The type of measure has two dimensions: the linkage to the objective, and the type of the scale. A "direct metric" is a measure directly linked to the objective, such as the number of nuclear missiles dismantled. Analysts and managers prefer to use a direct metric because it directly reflects progress toward the objective

[23] The committee offers general advice for all of the CTR programs, but CBEP and WMD-PPP are open-ended, expanding capacity-building programs and the committee has additional specific advice for them.

that a program is trying to achieve. "Proxy measures" are indirect measures used when direct metrics are not possible (cannot be measured) or are unavailable, such as when the direct metric is not timely. For example, cholesterol level is a proxy measure for the risk of heart disease. A proxy measure typically does not encompass all of the important aspects of status or performance, so managers sometimes use multiple proxy measures. Multiple proxies can increase the data collection requirements and only provide marginal capability to assess the achievement of the objective, but they may be all that are available.

The type of measure can be natural or constructed. A natural measure is a metric that has a commonly accepted definition arising from an objective measurement. Temperature and weight are examples of natural scales, as are silos destroyed and chemical destruction facility availability. Natural scales are preferred because people understand them intuitively and are easier to measure. Constructed scales are designed to measure the achievement of an objective. The five star scale used to assess the quality of a product, security classification levels (unclassified, confidential, secret), heat index, and body mass index are all constructed scales. Constructed scales require clear definitions to be operational to assessors and managers.

BOX 3-1 Common Terms Used for Metrics

- Cost effectiveness: Assesses attainment of an objective or task relative to the costs.
- Benefit cost measure: A metric reflecting benefits attained relative to the costs, which may be monetary or other costs.
- Value versus costs – Compares the value aligned with our objectives to the costs.
- Value measure – Align with our objectives, which we value.
- Measure of effectiveness – Assesses achievement of an assigned objective or task.
- Measure of merit – Another term that tells us how we achieve something we care about.
- Outcome measure – Focuses on the ultimate intended or unintended results.
- Performance measure: Describes how well a subsystem, system, or process meets its required performance.
- Output measure – Outputs usually refer to direct results of a process.
- Efficiency measure: A measure of how resources used in a project or program.
- Process measure – Captures some attribute of a process.
- Input measure – Identifies the resources or activities provided prior to a process.
- Resource measure: Identifies the resources (e.g., dollars, people's time, or materials) used by system or process.
- Leading indicators: Measures that provide early indicators of impact and effectiveness and allow managers to take corrective action on the process if required.
- Lagging indicator: A measure that is available after the current activity allowing corrective action only for future processes and operations.
- Environmental measure: A factor that could have a direct or indirect impact on the system or process but is not under the control of the program or project managers.
- Adversary measures: Assesses the potential or actual actions of adversaries that can have a direct impact on the process or system.
- Threat measures: Assess capabilities and intent of a potential opponent
- Criteria – Factors that differentiate alternatives.
- Attribute – Characteristic of a system or process.
- Metric – A standard of measurement.

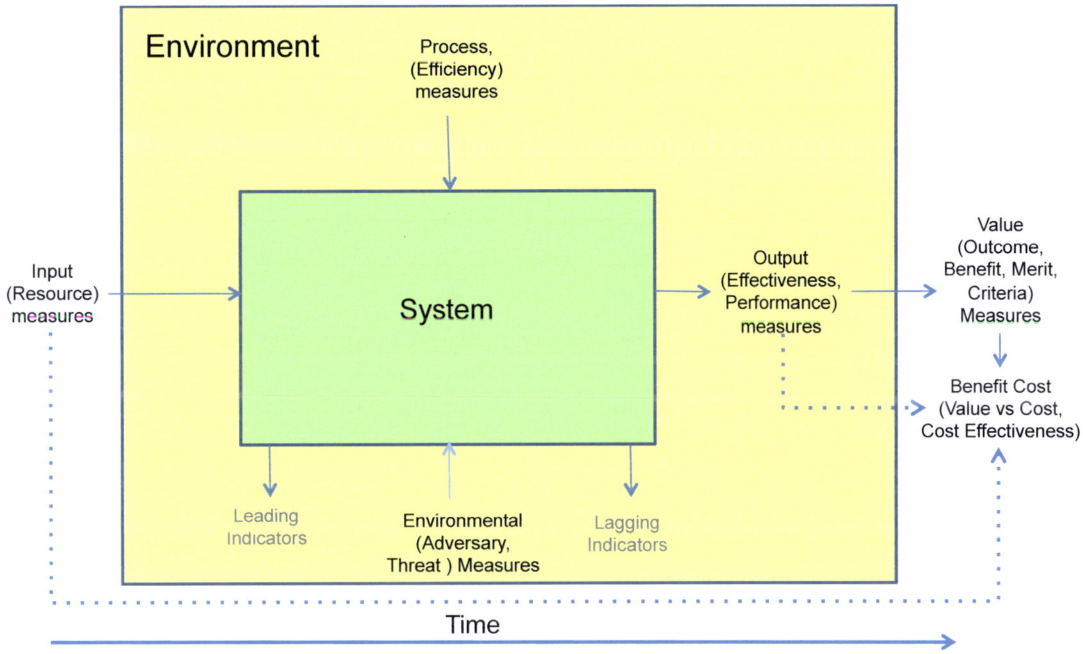

FIGURE 3-1 Metrics organized into six categories using systems thinking. The term on the top of each category is the term the committee uses in the report to assess the CTR metrics. Depending on the metrics' timeliness it could also be assessed as a leading or lagging metric.

For each objective, the Department of Defense's (DoD) best option is to identify a natural, direct metric. If that is not possible, the objective may be decomposed into tasks that are more measureable. If DoD cannot identify a natural, direct metric then DoD should develop a constructed scale metric. Because proxy measures only capture part of the objective, proxy measures should only be used if decomposition and constructed scales are not practical (see Table 3-1). There are other important qualities that make a metric more or less attractive: validity, reliability, sensitivity to change from interventions, cost of collection, and overlap with other measures of objectives.

TABLE 3-1 Classification of metrics by alignment with objective and type of scale ranked according to preference (1 to 4). The first preference is a direct, natural measure because it focuses directly on the objective and is well understood. The second preference is a direct, constructed measure. It is more important to be directly focused on the objective even if managers have to design a constructed scale. The third preference is a natural proxy measure, and the last preference is a constructed, proxy measure.

Type of Scale	Alignment with objective: Direct	Alignment with objective: Proxy
Natural	1	3
Constructed	2	4

What Metrics Cannot Do

The committee already described the important role of metrics in strategic (program) planning and tactical (project) implementation. However, it is also important to emphasize that metrics are a management tool and are not an end in themselves. There is some information that metrics cannot tell us, and there are several metrics implementation challenges.

First, metrics are only as good as the strategic thinking devoted to their development. For complex problems there is always the potential that an important objective was omitted when metrics were developed. Because metrics should follow from and be aligned with objectives, weaknesses or oversights in strategic thinking may lead to deficiencies in metrics. Stated another way, no set of metrics, no matter how elaborate or sophisticated, can compensate for a flawed strategy; hence the necessity of a sound strategic underpinning and well-formulated objectives. The best guarantee that one has a full set of metrics is meaningful discussion with *all* key stakeholders including the strategic and tactical leaders in the host country.

One aspect of strategy development that often requires special attention or emphasis is the identification of alternatives. There are usually multiple options for pursuing a given set of objectives, each with its own advantages and disadvantages. Arguments about metrics sometimes reflect more fundamental disagreements about how best to solve a problem, not how to measure it. Although a set of metrics can measure the impact and effectiveness of a project or program against objectives, and can even be used to compare competing options in some cases, metrics alone cannot ensure that the best option or options have been identified and are being implemented.

Second, metrics do not necessarily or automatically identify the cause of an adverse trend on a metric. This is especially true for output metrics. Managers and overseers may need a separate investigation, perhaps using specialized measures of the contributing causes to events or outputs to identify and understand the root cause of a metric trend. Following these trends can also lead to the observation of "unintended consequences," which may be positive or negative but worthy of note.

Third, metrics do not identify the best corrective action when a metric trend is not according to plan. Focusing on the root cause of an adverse trend on a metric may require a new objective and one or more new metrics. Unfortunately, if the metric is a lagging metric it may be too late for effective corrective action. It may, however, still be useful in future planning.

Fourth, it is important to remember that metrics are dynamic. The initial metrics may be well suited to the formulation and initiation stages of a program but may not adequately account for issues and concerns that arise later during program implementation. Therefore, it is important that the objectives and the metrics be dynamically updated throughout the project lifecycle, while avoiding the problem of "moving the goalpost."

Fifth, metrics cannot measure something that does not happen. Deterrence and some kinds of prevention are in some sense unknowable and unmeasureable. Proxy measures may be possible in these cases. The U.S. Government has a model of what leads to or exhibits risk. That model may be explicit (a stated set of risk factors that can lead to national security risks for the United States) or implicit (unstated assumptions about risk factors), but same model forms the basis for program strategy and planning. Proxy measures can be derived from the model, but the most desired outcomes (no WMD attacks) cannot be shown to follow directly from any particular actions.

This is why in assessing the overall effectiveness of the metrics for a program like CBEP, DoD has to step back from the numbers and ask itself whether the metrics reveal if the program is succeeding in establishing the human and institutional relationships, the interactions, and the transparency that result in trust and confidence between and among the partners.

Measuring Trust, Confidence, and Goodwill

How does one measure intangible impacts such as goodwill, mutual respect, collegiality, partnership, trust, and confidence? These are some of the objectives of scientist engagement programs, and they seem to be goals for which anecdotal rather than hard quantitative measures might be appropriate.

In some fields of endeavor, surveys are crafted to learn respondents' opinions, and conclusions are drawn from sufficiently large samples. For example, an extensive study of Russian WMD scientists was conducted to determine if respondents would be tempted by offers of WMD employment in rogue states (Ball et al., 2004). In the absence of International Science and Technology Center assistance, the risk of scientists "going rogue" was assessed as being significantly greater than with the assistance firmly in place. In the words of the study's authors:

> Our data from an unprecedented survey of 602 Russian physicists, biologists, and chemists suggest that the threat of WMD brain drain from Russia should still be at the forefront of our attention. Roughly 20 percent of Russian physicists, biologists, and chemists say they would consider working in rogue nations such as North Korea, Iran, Syria, or Iraq (still considered a rogue state at the time of the survey). At the same time, the data reveal that U.S. and Western nonproliferation assistance programs work. They significantly reduce the likelihood that Russian scientists would consider working in these countries. (Ball et al., 2004)

Surveys may be helpful in CTR programs. At the same time, DoD should be careful in relying on surveys, as there can be volunteer or self-selection bias (who participates) and response bias (deliberate or inadvertent skewing of responses to conform to an expectation rather than reflecting true beliefs), and a variety of other potential problems.

Another study (Revill, 2009) used surveys and other non-traditional measures to assess the effectiveness of dual-use educational efforts. In that study, concepts from the social sciences are proposed to augment the toolkit of traditional metrics. The author concludes that a multi-pronged approach to assessment seems the most likely to yield useful actionable results, at least in their field of endeavor. They note:

> With these points in mind, it is possible to identify a number of traditional and less orthodox methods of evaluation. However, as standalone measures none is likely to be successful in generating a holistic evaluation of any programme or process. Rather it is likely that a mixed method will be required to evaluate dual-use education that blends a number of different methods including inter alia, Questionnaire/Surveys; Likert model questionnaires; Social Network Analysis Content Analysis and Impact Evaluations. Using these combinations of measures it is argued that there is great scope for assessing dual-use educational objectives during and after a project and thus providing a much clearer understanding of what the process has achieved and how it could be improved in the future.

Ultimately, one might simply take responsible, reliable, and collegial behavior to be an indicator of outcomes (i.e., did the scientist contact international colleagues when dangerous infections or anomalies arose? Are the laboratory operations transparent and comfortable with international collaboration?).

RECOMMENDATIONS TO IMPROVE METRICS

The committee has several recommendations for how DoD can improve its metrics for the CTR programs.

Objectives and Partnership

> **For each program in the DoD Metrics Report, DoD should include a concise statement of its objectives and of how the program is intended to reduce threat or risk.**
>
> **Objectives for projects and the overall CTR Program in a partner country are developed jointly between the United States and the partner country. An agreed set of metrics should also be built into projects from the outset. They may change, but the parties responsible for the projects should know at any given time the metrics that will be used to measure impact and effectiveness.**

The CTR Program was established by Congress with clear authorities and each activity must begin with a clear statement of the United States' authorized overall objectives. DoD then needs to work with partner countries to define mutual, high-level objectives for their joint efforts. To measure impacts and effectiveness, metrics must include output metrics (e.g., changes in interdiction rates at borders) not just input metrics (e.g., training materials provided). DoD should identify the capabilities it needs to achieve its high-level objectives, the objectives for each capability, and where possible develop its metrics from outputs linked to the capacities that the programs are trying to build. DoD should also build the program and capability objectives into the DoD-partner agreement, and include provisions for metrics and the means to carry out those measurements.

As DoD takes CTR to new countries, it has opportunities to utilize lessons from 20 years of experience with cooperative threat reduction. The committee summarizes a logical order to developing metrics as follows.

1. State clearly the high-level objectives of the overall U.S. CTR Program, including linkage to threat or risk.
2. Work with the partner country to define high-level objectives for joint activities in their country. U.S. goals and partner-country goals need not match exactly, but they must be compatible.
3. Identify the capabilities needed to achieve the high-level objectives.
4. Define objectives for each capability being developed.
5. Define metrics with the partner country at the outset based on capability objectives. Agree on baseline, milestones, and measures of success. Different metrics may be appropriate for different stages of the program.

6. Build metrics—including exercises, if appropriate—into the implementation of the program.
7. Evaluate results independently (U.S. only) and together with partner country.
8. Feed evaluation back into decision-making process. If circumstances or other factors lead to a change in objectives, the metrics may need to be revised.

Working with and Learning from Other Agencies

> **DoD plans to leverage other U.S. Government agencies' experience, capabilities, and assets as CTR expands to new countries and as it continues existing programs. DoD also needs to communicate, coordinate, and cooperate with relevant agencies.**

As noted in Chapter 2, DoD plans to leverage other U.S. Government agencies' experience, capabilities, and assets as CTR expands to new countries and as it continues existing programs. DoD also needs to communicate, coordinate, and cooperate with relevant agencies.

One might argue that cooperation should be measured implicitly by looking at outcomes. Unlike safety and security (discussed in regard to the Chemical Weapons Elimination Program in Chapter 2), cooperation itself is an input and an outcome. So, why create metrics for cooperation if cooperation is an input measure (means) rather than an outcome (an end in itself)? It can be difficult to link interagency cooperation to a particular outcome in CTR, but there is experience indicating that problems arise from lack of coordination and lack of cooperation. It stands to reason that an agency newly entering a partner country will be more effective in that country if it learns from agencies that already work there. Furthermore, cooperation can serve as a leading indicator (as long as it is not simply a checkbox metric)—by the time an outcome measure indicates a problem, it may be too late to correct the lack of cooperation. The committee has struggled with this problem: A whole-of-government approach is critical to the success of some programs, and especially some planned programs, so a metric for interagency consultation, coordination, and cooperation would be a useful indicator for the implementation of a program. But the committee has not identified or developed a metric that is more than a checklist. Therefore, the committee encourages DoD, CTR experts outside of DoD, and experts in program assessment and evaluation to consider this challenge, but the committee makes no recommendation on a metric for taking a whole-of-government approach, important as it may be.

DoD is not the only agency engaged in capacity-building programs. U.S. Customs and Border Protection conducts a mission similar to the PPP along many thousands of kilometers of border and has developed metrics for its mission and operations. The United States Department of Agriculture's Animal and Plant Health Inspection Service is a leader in an international surveillance network that has many parallels to the global network DoD leadership envisions for biothreats. United States Agency for International Development operates programs with partners across the world to foster democratic institutions, compared to which DoD's capacity-building programs seem relatively tangible. The DoD Defense Security Cooperation program also shares important similarities with the DoD CTR Program and may in some cases serve as a model. Not only can DoD learn from other agencies, but DoD will be working with other agencies in a "whole of government" effort, and they might already have mechanisms in place for measuring impact and effectiveness that would be useful to DoD CTR.

From the variety of the national security programs that have comparable metrics challenges to the CTR Program, the committee selected three of these programs for comparison to gain insights into the CTR metrics approach. The Defense Security Cooperation Program, Capacity Building in Iraq and Afghanistan, and the DoD Capabilities Based Planning Program each illustrate important aspects of CTR. Table 3-2 displays several features of the four programs: the program purpose, scope, number of programs, objectives, metrics, and use of metrics.

The Defense Security Cooperation Program uses metrics to assess security cooperation gaps and the contributions of projects in each of the geographic regions assigned to each Combatant Command (COCOM) under the Unified Command Plan. In addition, many of the supporting commands and services (e.g., U. S. Army Pacific Command and Marines) use metrics to assess the contributions of their security cooperation programs to command missions and security objectives.

> The Defense Security Cooperation Agency serves as the DoD focal point … for the development and implementation of security assistance plans and programs, monitoring major weapon sales and technology transfer issues, budgetary and financial arrangements, legislative initiatives and activities, and policy and other security assistance matters. (DoD Directive, 5105.65)

Capacity Building in Iraq and Afghanistan is a complex process that uses metrics to assess the progress of the approved Campaign Plan and to assess the progress of individual projects that build specific infrastructure capabilities in specific locations. The typical approach

TABLE 3-2 Comparison of the CTR Program with other DoD programs that face challenges developing metrics

Program Feature	Cooperative Threat Reduction	Defense Security Cooperation	Capacity Building in Iraq and Afghanistan	DoD Capabilities Based Planning
Purpose	Reduce nuclear, biological, and chemical WMD risk to U.S. national interests	Increase security cooperation to support U.S. national interests	Build the infrastructure to provide for security and economic development	Develop future U.S. military capabilities
Scope	CTR Program objectives	Defense and COCOMs regional objectives	Mission assigned by President and Commander's Campaign plan	Capability gaps are identified in defense mission areas
Sub-Programs	Several projects for each major program	Many projects in each COCOM	Many projects to develop country capacity	Capability gaps determine need for weapons systems
Objectives	CTR and project objectives	COCOM regional security objectives	Commander's objectives	Acquisition program objectives
Metrics	Project only	Metrics for each region and country	Metrics for each objective	Key Performance Parameters and Key System Attributes
Use of Metrics	Evaluate progress of the 4 annual projects	Identify gaps and assess annual projects	Assess campaign progress and project progress	Assess program progress

to metrics is to develop a hierarchy of missions, objectives, tasks, and metrics. While quantitative data are obtained on each of the metrics, typical reporting to senior leaders may use red, yellow, and green colors to simplify the quick identification of gaps and problem areas (where red means the objectives have not been met, yellow means they have partially been met, and green means they have been or will soon be fully met). Due to the long time horizons to build national capacity, the operational analysts that perform and present the data to senior leaders report a tension between the focus of the project leaders who are interested in reporting on the successful completion of their short-term projects and the campaign planners who must assess the long-term impact and effectiveness against campaign objectives.

The DoD capabilities based planning process replaced the threat based planning process used in the Cold War. Paul Davis defined capabilities based planning as "planning under uncertainty to provide capabilities suitable for a wide range of modern-day challenges and circumstances while working within an economic framework that necessitates choice (Davis, 2002). The process has grown in complexity but the basic ideas are straightforward. The process begins by assessing each mission area to define the joint capabilities that will be needed in the future. This typically is done using a hierarchy of capabilities, objectives, tasks, and metrics. The next step is a capability gap analysis to identify the capability gaps for each mission. After the capability gaps are identified, a functional solution analysis identifies and evaluates joint concepts to provide the needed capability. Finally, where necessary, acquisition programs are initiated to obtain the capability to fill the gap. A capability development document defines the program success metrics which are called Key Performance Parameters with minimum performance (threshold) and aspirational goals (objective levels) and Key System Attributes.

Comparing the CTR Program with similar programs provides insights into the different uses of metrics. One of the major differences between the CTR Program and the other three programs is that the CTR metrics do not address the full scope of the threat in each of the WMD areas. Instead, the metrics address only the scope of the funded projects. While the project metrics are useful in assessing the annual project status, they do not help Congress and senior leaders in DoD and other parts of the government to understand the full scope of the potential for cooperative threat reduction which could help identify the need for and scope of future projects.

Observations from Non-Security Programs

Across areas as diverse as public health, poverty reduction, democracy promotion, public education and others, experts have been attempting for years to employ systematic efforts to understand the effectiveness of their programs. Cultural sensitivities associated with many of these initiatives, constraints of limited resources as well as a sincere interest in positive change have motivated multiple and sustained efforts to develop effective metrics for these programs. Their task is made more complicated by the fact that their objectives – improving public health, reducing poverty, developing democracy, and increasing educational achievement – are at least as difficult to measure as those of capacity building and improving cooperation through CTR projects. Despite the challenge of trying to measure objectives difficult to quantify, much progress has been made in developing methodologies and metrics to do just that. A few examples may provide useful lessons for the DoD CTR programs.

The field of global public health assistance (see e.g., Unite for Sight, 2010) has spent considerable resources on healthcare efforts—as much as $14 billion in 2004 (Kates et al., 2004)—and therefore donors as well as communities affected have increasingly demanded better understanding of results of those investments. This has led to the explicit differentiation between outputs (goods or services generated by the programs) and outcomes (impacts of programs). For management and contracting purposes, it is important to track outputs, but far more important is measuring outcomes, which in the public health area as well as other areas, include "changes in behavior, attitude, skills, knowledge or condition" (Unite for Sight, 2010; see McAllister, 1999).

How does one measure attitudinal and behavioral change that would ensure that the desired outcomes are achieved? Stating the objectives as part of the initial program design by focusing on outcomes over outputs allows for the development of metrics that, while perhaps more difficult to measure, gets closer to understanding the extent to which the program actually reaches the stated goals.[24] The global health assistance community has been working to further develop these methodologies and metrics. Specific efforts include those of the Health Metrics Network and the Institute for Health Metrics and Evaluation. The World Health Organization and International Monetary Fund also continue to work on the challenge of metrics development necessary for understanding outcomes/impact.

Another equally challenging area to measure is that of democracy assistance, beginning with the definition of "democracy" itself. Attempts to better understand the considerable challenge of measuring such a difficult object as "improving democracy" led the U.S. Agency for International Development to ask National Academy of Sciences to bring social science methodologies to bear on the challenge. The report resulting from this request concluded that due to the difficulty of the overall challenge, a host of methodologies is needed to more effectively conduct impact evaluations, defined as evaluations that assess whether and how the specific program made an impact in the expected (or even unexpected) ways. In other words, the methodologies for evaluating the impact of projects are designed to understand what would have happened if the USAID programs did *not* occur. The methodologies proposed in the 2008 report include case studies, systematic interviews, and randomized studies (NRC, 2008).

Perhaps more important than the adoption of a particular methodology is the selection of a rigorous methodology designed from the outset to measure progress toward the objectives of the program along with its impacts and effectiveness.[25] In the USAID context, there are three types of objectives: project monitoring (routine oversight or project management), project evaluation (is the project having its intended affect), and country assessment (is the country as a whole improving). Each of these objectives requires different metrics that specifically and systematically match these tasks with corresponding measures.

Equally important is the collection of appropriate data for each of these measures both before the project begins (baseline data) and after the project is completed (outcome data). The same data should be collected wherever possible for comparable individuals, groups, or communities that, whether by assignment or for other reasons, did not participate in the program. (NRC, 2008). The 2008 report concludes that determining effective metrics for each of the

---

[24] Moreover, public health programs require a considerable degree of cooperation and integration with local government and community leaders to establish joint goals, methods of implementation, and measurement of outcomes/impact. Learning more about these interactions may provide helpful insights for DoD since there is important overlap between the ability of a public health system to benefit the public overall as well as to accomplish security objectives.

[25] What is the difference between progress and impacts? Consider the Chemical Weapons Elimination program. Years were spent planning and constructing the chemical weapons destruction facility. Although the program may be making good progress, that part of the program has little impact until it begins operations actually destroying chemical agent.

objectives and the collection of baseline and outcome data will lay the groundwork for appropriate impact evaluations of USAID's programs.

Finally, however, they found that "even if USAID were to complete a series of rigorous evaluations with ideal data and obtained valuable conclusions regarding the effectiveness of its projects, these results would be of negligible value if they were not disseminated throughout the organization in a way that led to substantial learning and were not used as inputs to planning and implementation of future [democracy and governance] projects" (NRC, 2008). This same learning process would be useful to DoD as it evaluates the impact of its CTR programs.

Prioritization

The committee judges that using a consistent framework to prioritize and refine metrics within a program would help DoD and other CTR decision makers. Using such a framework, DoD can identify the highest priority metrics, ensuring that the metrics are useable and useful, and allow decision makers to feed results back into the overall CTR objectives and budgetary process. Any of several decision-making or prioritizing frameworks would work, including the decision analysis technique of swing-weight analysis and the DoD capabilities based planning process.

As noted in Chapter 2, DoD developed each set of metrics *ab initio*, using a different approach for each program, i.e. chemical, biological, borders, and nuclear. The committee found it difficult to identify what were the most important metrics and what might be missing from DoD's metrics because DoD used a different approach for each program. Although each CTR program is distinct and possibly unique, the committee's suggestion that DoD consider using a consistent framework is based on best practices, the fact that using a consistent framework is easier, and the committee's hopes that DoD will maximize what it can accomplish with metrics. Using a well-developed framework helps ensure completeness (Did we include everything that matters?), internal consistency (Did we double count? Do these pieces fit together logically?), and focus (Did we remove what doesn't matter?). It also helps with management and oversight because it offers the possibility that managers might be able to compare across programs, and does not require the managers and overseers to learn a new framework for each program. The burden then is on those who prefer using different frameworks to argue why they should be different.

An important component of measuring what matters is prioritizing among capabilities and within sets of metrics. Not all goals are equally important and the act of prioritization will help focus limited resources. DoD should use a consistent framework to prioritize and refine metrics within program, for both the decision and the prioritization frameworks. In this report, the committee highlights the decision analysis technique called swing-weight analysis (described below) and the framework used widely in DoD called capabilities based planning.

Using a consistent framework does not mean that the metrics for each CTR program or program activity should be the same, but that a common framework should be used for defining objectives and metrics should flow from those objectives. This will make the metrics useable and useful, and allow decision makers to feed results back into the objectives and budgetary process. Prioritization is a critical component of the effective and efficient use of metrics in program assessment and management.[26] Prioritization is also an essential part of the capabilities based planning process. The objectives and the capabilities needed to meet those objectives are prioritized.

A key technique for prioritizing metrics is the swing weight matrix (Parnell et al., 2011), which can and has been used as the prioritization tool in a capabilities based planning exercise. A swing weight matrix defines the importance and range of variation for a set of metrics. The idea of the swing weight matrix is straightforward: a metric that is very important to the decision should be weighted higher than a metric that is less important. A metric that differentiates between alternative outcomes is weighted more than a metric that does not differentiate between alternative outcomes (see Box 3-2).

BOX 3-2 Swing Weights, A Simple Example

Decision makers use swing weights to identify the most important metrics for a decision (Keeney and Raiffa, 1976). Swing weights depend on importance and range of variation for the metric. The importance is a subjective judgment and the variation in the metric is a factual assessment. For example, consider a car buyer who plans to select the car based on *n* factors or car metrics. Many people would subjectively judge that car safety is an important metric in the decision to purchase a car because they value the safety of the passengers (themselves, their family, and their friends). Using a website that summarizes car features, the buyer finds that each car receives a rating on a 5 star safety scale, where a one-star car is in the bottom category, meeting only minimal safety requirements for crashes and rollover events, and each additional star represents better safety. A five-star car performs well on all of the crash and rollover tests. Consider the following sequence of judgments and the resulting impact on the swing weights.

1. Suppose that the buyer is willing to consider any car that has a safety rating of 1 to 5 stars. Safety is very important and the metric variation is large. Therefore, the metric might be assigned a high weight, say 20%. The other $n-1$ metrics would have a total weight of the remaining 80%.
2. After some thought, suppose the buyer decides to consider only cars with a safety rating of 3 to 5 stars. Safety is still very important but the metric variation has been significantly reduced. Therefore, one might assign a lower weight, say 10%. The other $n-1$ metrics would have a total weight of 90%. The extra 10% would be spread by the same relative proportions of the weights of the $n-1$ metrics.
3. After rethinking again, suppose the buyer decides to consider only cars with a safety rating of 4 to 5 stars. Safety is still very important but the metric variation has been reduced. Therefore, the buyer might assign a lower weight, say 5%. The other $n-1$ metrics would have a total weight of 95%. The extra 5% would be spread by the same relative proportions of the weights of the $n-1$ metrics.
4. Finally, suppose the buyer decides to consider only cars with a safety rating of 5 stars. Because all cars under consideration have the same safety rating, safety is no longer a metric that distinguishes among the alternatives. Therefore, the buyer would assign a weight of zero. The other $n-1$ metrics will have a total weight of 100%.

[26] Large programs typically have many metrics. The Pareto Principle or the 80/20 Rule applies to metrics. The 80/20 rule was initially developed by economist Pareto in 1906 when he noted that about 20 percent of the people owned 80 percent of the wealth in Italy. Juran, a U.S. quality expert, extended the Pareto rule to management in the 1930s. For example, many project managers have observed that about 20 percent of the tasks on a project require about 80 percent of the time and resources. Likewise, not all metrics are equally important and some metrics may not be significant enough to spend the time and effort to collect and analyze data and report results. Applying the 80/20 rule to metrics, we would expect that about 80 percent of the importance is in about 20 percent of the metrics.

The first step is to create a matrix (for example, Table 3-3) in which the top row defines the metric importance and the left side represents the range of variation. The levels of importance and variation can be thought of as constructed scales that have sufficient clarity to allow program managers to uniquely place every metric in one of the cells. Continuing from the discussion of priorities in Chapter 2, an illustrative swing weight matrix for the CBEP program is displayed in Table 3-3. Using the swing weight matrix to prioritize the metrics, the importance is defined in three levels, with direct reduction in threat being the most important. The vertical scale defines four levels of the gap between the desired capability and the current capability.

TABLE 3-3 An illustrative swing weight matrix for the CBEP program.

		Importance, f_i, of the metric i (on a scale from low = 1 to a high = 100)[a]		
		Direct reduction in the biological threat	Demonstrated use of biosafety and biosecurity procedures	Development of biosafety and biosecurity plans and procedures
Variation in the capability due to the range of the metric	Large capability gap	Consolidate EDP[b] - 100 Secure EDP– 100	Copies of EDP strains sent to United States – 50	Major biosafety and biosecurity plans – 10
	Medium capability gap	Budget sustainment resources – 50	Demonstrated EDP detection and timely reporting – 25	Other important plans – 5
	Small capability	Research programs aligned with national & international EDP priorities – 10	Other detection and timely reporting – 5	All other plans and procedures – 1
	None	Eliminate known biological weapons (assume none in the 6 partner countries)		

[a] EDP = especially dangerous pathogen

[b] There may be more than one metric with a weight of 100.

The swing weights are assessed by assigning the most important metric an arbitrary value, e.g., 100. The rest of the weights are then assessed relative to the most important metric or any assessed metric. Table 3-3 provides an illustrative example of a possible swing weight assessment. The metric weights can easily be normalized to sum to 100 using the following equation.

$$w_i = \frac{f_i}{\sum_{j=1}^{n} f_j}$$

In this illustrative case there are n = 10 metrics, and each metric i would have a normalized weight w_i.[27] Additional details on the mathematical constraints on the swing weights (decrease as

[27] As it is described here, swing weight can help to prioritize and assign importance weights to individual metrics. If the metrics are both quantitative and intercomparable on the same scale using a performance function, then once those weights are established, a summary score can be calculated by multiplying the graded performance score on each metric with the metric's importance weight, then summing up all the weighted scores to get an overall measure of aggregate performance. However, even if the metrics are quantitative, mixing them together might be invalid or just misleading (for example, the metrics might be nonlinear or interdependent). The mathematical assumption is that the measures are mutually preference independent (Kirkwood, 1997).

you go down and to the right) and techniques for assessing the weights and performing sensitivity analysis can be found in Parnell et al. (20ll).[28]

Border Security Metrics

Law enforcement agencies, whether the U.S. Border Patrol or local police, have a long tradition of collecting and using statistics to measure various aspects of their operations and performance. When combined with the judgment and experience of trained professionals, such data can be and have been used effectively to focus resources, communicate priorities, and identify and adjust to changing conditions, such as changes in the patterns of illegal activity or evolution in the capabilities or behavior of criminals or illegal border crossers.

For example, for border security, a classic metric is the number of illegal alien apprehensions per year (or month or quarter).[29] Although data supporting this metric are routinely collected and easily compiled, it is widely understood that interpreting the metric is not simple because it reflects not only effectiveness of detection and interdiction (the rate of apprehensions, given the flow of illegal crossings), but also the highly dynamic nature of cross-border traffic, which depends on many factors such as relative economic conditions on both sides of the border, seasonal patterns, tactics of the illegal crossers and their facilitators, and the perceived risks of apprehension (including both the chance of being caught and the consequences if it happens—ranging from "catch and release," to temporary detention, longer-term imprisonment, deportation to various points in the country of origin, and so forth).

Nevertheless, the metric can be informative if carefully interpreted. The number of apprehensions by the U.S. Border Patrol has declined significantly over the past decade, from a high of 1.7 million in fiscal year 2000, to 463,362 in 2010, a remarkable decline that may reflect a significant improvement, if one can determine that a greater decline was not possible. In other words, it is reasonable to judge this decline to represent an improvcment, but this cannot be known for certain, nor can the exact reasons for the decline (economic recession in the United States versus increased risks of failure versus changed tactics, for example) be known with any precision. More importantly, while the number (or rate) of illegal crossers who are *not* apprehended, which is of key concern to many stakeholders, cannot be measured precisely, it can in fact be estimated through analysis of know evasions, investigative intelligence, interior enforcement actions, and other poll data regarding the number of illegal aliens present in the United States. Making that estimate is important to understand whether a decline in apprehensions is actually indicative of successful tactics. Further, while this estimation is difficult in the United States, it may be even more difficult elsewhere.

Other easily collected and compiled border security metrics include:

- miles of pedestrian and vehicle fencing
- number of border patrol agents
- numbers of various classes of equipment deployed such as unmanned aircraft systems, remote video surveillance systems, mobile surveillance systems, unattended ground sensors

[28] See Parnell et al. (2011).
[29] See www.cbp.gov/linkhandler/cgov/border_security/border_patrol/uslop_statistics/total_apps_25_10.ctt/total_apps_25_10.pdf; accessed November 11, 2011.

- quantities of contraband seized (pounds of marijuana, cocaine, heroin, methamphetamine, numbers of weapons, amounts of currency, etc.)
- incidents of violence against border patrol agents
- percentage of time detection machines are operational

Similar kinds of data are collected by CTR partner agencies in the WMD-PPP. Again, while useful and informative, such statistics measure activities and to some extent outputs, but not overall effectiveness or results. Recognizing the limitations of these kinds of traditional statistics, beginning in 2004, the U.S. Border Patrol formulated an alternative metric to supplement the traditional ones: "miles of border under operational control" (see GAO, 2011). This was a constructed metric that relied on integral judgments by local border patrol officials about the level of security in their area, using a five-level scale. The five levels were:

1. controlled: continuous detection and interdiction resources at the immediate border with high probability of apprehension upon entry
2. managed: multi-tiered detection and interdiction resources in place to fully implement the border control strategy with high probability of apprehension after entry
3. monitored: substantial detection resources in place, but accessibility and resources continue to affect ability to respond
4. low-level monitored: some knowledge is available to develop a rudimentary border control strategy, but the area remains vulnerable because of inaccessibility or limited resource availability
5. remote/low activity: information is lacking to develop a meaningful border control strategy because of inaccessibility or lack of resources (see GAO, 2011)

A recent assessment based on these definitions and on information supplied by the border patrol found 873 miles of the southwestern U.S. border were under "operational control," defined as either "controlled" or "managed." Of these 873 miles, 129 were "controlled" and 744 were "managed," and these figures showed significant improvement over past years. The remaining 1,120 miles of southwestern U.S. border were found to be not under operational control (GAO, 2011).

Although the operational-control metric comes closer to measuring results or effectiveness than the more traditional measures, it still has shortcomings. As the definitions indicate, it is somewhat subjective and combines aspects of "resources in place," which is an input, and probability of apprehension, which is an output. The probability of apprehension cannot be measured directly, although it might be inferred with some uncertainty from other data and exercises.

In addition, questions have been raised about whether the "operational control" measure properly reflects the overall objectives of border security. In Congressional testimony, Border Patrol Chief Michael J. Fisher explained: "Operational control is the ability to detect, identify, classify and then respond to and resolve illegal entries along our U.S. Borders." He further cautioned: "The term is tactical in nature and by current use can only be achieved by incrementally applying resources to a point where field commanders can consistently respond to and resolve illegal entries. Operational as a measure does not accurately incorporate the efforts of [Customs and Border Protection] partners and the significance of information and intelligence in an increasingly joint and integrated operating environment. The Border Patrol is currently taking steps to replace this outdated measure with performance metrics that more accurately depict the state of border security" (Fisher, 2011). This approach is built on a perception that as resources increase, the cost of getting caught also increases, therefore, the illegal entries decrease.

U.S. Customs and Board Protection Commissioner Alan Bersin in a different hearing said: "The success of our efforts … must be measured in terms of the overall security and quality of life of the border region; the promotion and facilitation of trade and travel; and the success of our partnerships in enhancing security and efficiency" (Bersin, 2011).

This debate, which will undoubtedly continue in Congress and elsewhere in the years to come, is a reminder of the fundamental point the committee emphasizes elsewhere in this report. The debate is a reflection of the operational reality that an effective boarder security system cannot rely upon a single actor or point of failure. Specifically, to be useful, metrics must be linked to, and ideally derived from, a clearly defined and agreed set of fundamental goals and objectives. Where fundamental differences exist about objectives, metrics can reveal or illustrate those differences, but cannot not resolve them. However, through an iterative process focused both on objectives and metrics, clarity about metrics can potentially inform and shed light on debates about goals and objectives.

The DoD Metrics Report mentions that "one way to measure the effectiveness of enhancements objectively is through testing; that is, standardized exercises that can be conducted before and after enhancements are made to measure the impact of those enhancements." In addition, the Report presents a set of more detailed project-level metrics that are intended to capture a wide range of specific impacts or effects in relation to a detailed hierarchy of goals and objectives.

The committee recognizes the difficulty of establishing a compelling overall effectiveness metric. As the section on border security metrics suggests, the question remains difficult and controversial even for the U.S. Border Patrol. In the absence of an overall effectiveness metric, it may be necessary to settle for a set of activity-based and output-based metrics. However, the committee has several recommendations for improvement.

As discussed elsewhere in the report, the committee believes that tests and exercises, including the use of red teams if and where appropriate, have an essential role in measuring effectiveness of CTR programs. The committee agrees with the potential value of exercises for measuring the impact of border security cooperation and is aware that such exercises have been conducted (e.g., the annual Operation Sea Breeze joint training exercise on the Black Sea). Although care should be taken in interpreting the results of exercises, to guard against overly optimistic appraisals and the tendency to perform better during an exercise than during unplanned events, the committee recommends that the WMD-PPP add exercise-based metrics explicitly to its collection of metrics.

The metrics defined in terms of "miles of border enhanced" and "number of ports of entry enhanced" are reasonable input metrics that measure the delivery of tools to accomplish border security goals. There is also precedent for them in other CTR programs implemented by other U.S. agencies such as the Department of Energy Materials Protection, Control and Accounting Program (numbers of facilities/buildings/sites with rapid/comprehensive security upgrades) and the Second Line of Defense Program (number of border crossings equipped with radiation detection capabilities). However, these are numerator statistics. To put them in perspective, denominators should also be provided. For example, comparing these numbers to the total number of miles of border and the total number of ports of entry that were determined to be within the scope of the program. The committee recognizes that selecting the denominators raises additional questions, such as "how was the scope of the program determined?" For example, most programs will not seek to upgrade 100% of all the possible borders or ports of entry, but will instead focus on providing a "model" for some subset identified through a prioritization or a risk assessment process, leaving the partner country with the task of extending the model as appropriate to the remainder of the border. The committee recommends that this step be made explicit and visible in the metrics because it ties closely to the question of overall goals and objectives of the program. Although disagreements and controversies may arise, evaluation is best served by making those explicit and explaining the planning assumptions on which the program is founded, even though full consensus may be lacking.

Biological Engagement to Enhance Security

The CBEP metrics in the DoD Metrics Report represent many important aspects of the program and yet still have room for improvement. In meetings with the committee, DoD specifically asked the committee to provide advice on how to simplify the set of metrics for CBEP. Table 3-4 provides the committee's tally of the CBEP metrics (the DoD Metrics Report

TABLE 3-4 Assessment of the Cooperative Biological Engagement Program Metrics

	Partner Domestic Stability (Environment)	Partner Capability			Partner Outcome	Total
		Input	Process (Existence)	Output (Conditional)[a]		
1. Secure & Consolidate			1		1[b] (Consolidate EDP)	2
2. Improve bio security & safety standards			8	2	1[b] (Consolidate EDP)	11[c]
3. Detect, diagnose, & report			8	17		25
4. Sustain capabilities		2 (Budget)	1	1		4
5. Engage scientific & technical people			2	3	1 (Copies of EDP stains sent to U.S.)	6
6. Eliminate BW technologies					1 (Eliminate BW)	1
Total	0	2	20	23	4[c]	49[c]

[a] Output is conditional on the event occurring.
[b] Same measure.
[c] Includes one repeated metric.
EDP = especially dangerous pathogens

calls the CBEP metrics indicators; see Appendix B) using the measurement concepts presented in this report. Three features should be noted. First, three important direct outcome metrics (one is duplicated) were identified. Second, one important input metric (the partner country budget for the program) was identified. These metrics are useful leading indicators of sustainability. Third, the specification of the minimum performance (threshold, in capabilities based planning parlance) and aspirational goal (objective) for each measure appear to match those used in the DoD acquisition process (see Box 3-3).

However, there are several areas for simplification and improvement in the development of the CBEP metrics.

- The presented structure in the DoD Metrics Report that suggests moving from program objective to capability is too cumbersome because the capabilities being developed through a CBEP program may support multiple objectives. This causes metrics to be

Box 3-3 Acquisition and Metrics

DoD has conducted the CTR Program as an acquisition program: DoD contracts with companies to implement CTR plans. Some of these plans are for elimination of nuclear warhead delivery vehicles; some are for acquiring capabilities (facilities, technology, and training). In some ways, this is not so different from other acquisition programs in DoD. For example, it is easy to see the parallel between DoD contracting with a company to tear down a building on a military base and DoD contracting with a company to destroy missiles or silos. But even the capacity building programs have parallels in other acquisitions. When DoD acquires a weapons system, the department is not just buying a weapon. A weapon on its own has little value. What DoD really acquires is a capability, comprising the equipment and the ability to operate and maintain the equipment.

DoD has a well developed process for acquisition, which includes evaluations of whether the department has not only taken possession of the equipment, but also acquired the capability to operate and maintain the equipment. In DoD acquisition parlance, there are two major milestones beyond the baseline for such an acquisition: initial operational capability (IOC) and full operational capability (FOC). The acquisition process is described in the Defense Acquisition Guidebook (DoD, 2011) and the following concise definitions of IOC and FOC can be found in the Glossary of Defense Acquisition Acronyms & Terms (DoD, 2005).

> Initial Operational Capability (IOC)
> In general, attained when some units and/or organizations in the force structure scheduled to receive a system 1) have received it, and 2) have the ability to employ and maintain it. The specifics for any particular system IOC are defined in that system's Capability Development Document (CDD) and Capability Production Document (CPD).
>
> Full Operational Capability (FOC)
> In general, attained when all units and/or organizations in the force structure scheduled to receive a system 1) have received it, and 2) have the ability to employ and maintain it. The specifics for any particular system FOC are defined in that system's Capability Development Document and Capability Production Document.

The IOC and FOC are different for each context and attaining the IOC and FOC for a complex acquisition typically requires that minimum performance be met on a variety of metrics. In some cases, that performance is established by exercises.

repeated.[30] A better structure would be to use two steps. In the first step, DoD should identify the capabilities that are required to achieve the program objectives, i.e., what capabilities for capacity building. In the second step, goals are set for each capability and metrics toward achievement of the goals are identified.

- The CBEP has no benefit-cost metrics to assess the benefit or effectiveness the United States achieves for the resources invested in the program.
- There are no metrics of the environment. For example, perhaps it would be useful to assess the political, economic and social environments of a country in which CTR programs are being implemented. Such assessments may be beneficial in terms of understanding the potential for sustainability challenges. Another environmental metric could indicate how cooperative the partner country is.
- The metrics use 20 proxy metrics of the processes required for biological safety and security. This is a large number of proxy metrics compared to the three direct, outcome metrics. It would appear that two constructed, process metrics (safety and security) would be sufficient to encompass the information in the 20 proxy metrics.
- All of the 23 output metrics are conditional on the emergence of some disease outbreak, involving dangerous pathogens occurring and the partner country executing the correct procedure according to its plans. Again, these metrics provide the numerator but we do not know the denominator.

The committee's overall assessment is that the CBEP metrics could be reasonably assessed with about 11 metrics instead of the 49 metrics listed in the DoD Metrics Report. The 11 metrics would be: 2 input (budgets), 3 process (safety, security, and technical engagement), and 2 output (safety and security), and the 4 outcome measures.

Why are these numbers of metrics important? The total number is important because if there are too many, it is difficult to understand the net result. Prioritization helps with this problem, making it possible to recognize the bottom line. Indeed, as noted above, DoD officials asked the committee to help managers identify the "bottom line." The lower number after prioritization is manageable and better reflects what matters.

Independent Evaluation

Capacity building programs need independent evaluation of how the capabilities being built perform in action. This can be accomplished by several means, ranging from independent expert observations of routine operations to comprehensive exercises that test the full scope of capabilities. The level of effort can be tailored to the scope of the program, its resources, and its relative importance. DoD and its partners should build such independent evaluation into each project. The Defense Security Cooperation Program might be a good model for how to proceed.

Measuring progress in building capacity and effectiveness of programs to prevent low-frequency, high-consequence events is difficult. Assessment against standards and guidelines (e.g., is the partner's action plan for interdicted nuclear material consistent with the International

[30] See CBEP Objective 2: measures 1.1 and 2.1 are repeats of CBEP Objective 1: measures 1.1 and 2.1.

Atomic Energy Agency model action plan?) is one important component, but DoD and Congress care more about likely performance when the event occurs (i.e., how effectively can the partner implement the action plan?). Current thinking on evaluation suggests that the involvement of independent evaluations from the beginning and for all aspects of the evaluation, including data collection, is ideal. Independent evaluation establishes a degree of credibility that is hard to achieve by other means. Especially for capacity-building programs, some kind of independent evaluation is essential. Exercises are a good way to measure effectiveness and sustainment. Exercises can help measure both capability and performance in such programs. While exercises might be structured differently in different countries and for different projects, they should be built into the implementation and evaluation components of capacity-building projects and programs from the beginning, starting with a baseline evaluation and proceeding with midcourse and final or sustainment evaluations.

The kind of evaluation employed needs to be tailored to the scope of the program, its resources, and its relative importance. The evaluation might take the form of periodic expert observations of the project operations or the partner country's capacities. It might be an impromptu test of a randomly selected part of the system (e.g., a border protection system), or it might be an exercise of the system, such as was performed for CBEP in Georgia. Ideally, the exercises would be designed by an entity independent from the groups being tested. Fully independent evaluation is not feasible for some of the programs, but recognition of why independent evaluation is desired (to avoid unintentional and intentional bias and to establish credibility) can help guide implementation of evaluations where truly independent evaluation is not done.

Care must be taken with any metrics, but especially exercise-based metrics, to avoid perverse incentives and gaming of the system. These programs are at least as much about human resources and relationships as about facilities and equipment. Because communication is a critical element of success, evaluation requires more than just the exercise or observation; it requires discussion afterward between credible parties (in expertise, authority, and attitude) on both sides.

Box 3-4 An Example of Exercises as Part of the Metrics for CTR

DoD used an exercise-based of means to test the capability of technologies, systems, and support procured by the DoD for the IOC assessment of the disease surveillance system in countries of the former Soviet Union (FSU), such as Georgia, Kazakhstan and Uzbekistan supported by what was then the Biological Threat Reduction Program under CTR. The IOC assessment included exercises in the recognition of clinical disease in humans and animals in the field, the collection of samples, the transport of samples to the appropriate laboratories, diagnosis the disease, reporting of results nationally and internationally as appropriate and destruction or storage of the agents in a secure repository. Prior to the IOC assessment, table top and field exercises were used as nongraded training exercises to focus on gap analysis for individuals and the overall system. Although the DoD acquisition process establishes many of the requirements for the implementation of CTR programs, the DoD Metrics Report does not make clear the relationship between the current metrics and the DoD acquisition process. For example, it is not clear whether DoD will use the IOC and the Full Operational Capability (FOC) as the sets of metrics to determine CBEP's success. The DoD metrics used for the IOC assessments in the FSU countries appear to have been adopted for CBEP in the DoD Metrics Report. However, the initial application of the current metrics for CBEP yielded a different assessment from the IOC for one of the countries. If essentially the same metrics are used for the IOC and the evaluations based on the DoD Metrics Report, then one would expect that the results would be similar, absent any significant deviation in the performance of the host country.

If DoD continues to use the acquisition process for CTR programs, attainment of the IOC and FOC will serve as the main metrics of success or failure, so it would make sense to integrate the metrics used to evaluate attainment of the IOC and FOC with the formal metrics of the program. This would avoid competing assessments.

At the same time, there is some question whether acquisition is the right framework for some kinds of capacity building, such as bioengagement. Biological threat reduction began in the FSU where there had been an active biological weapons program, where the infrastructure, regulations, and training were similar, and the improvements by DoD in infrastructure and training were essentially the same. But the new mission and locations (Africa, Asia and South America) are different. While the overall objective may be the same, how DoD engages and how DoD supports these countries to meet the overall objective will likely be different. If relationships, trust, and a culture of responsibility are some of the outcomes CBEP seeks, is the acquisition model the best way to cultivate and measure those capacities? This question is worth DoD's consideration as it implements the program and its metrics.

Building in Exercises From the Beginning

For exercises to be successful, DoD needs an agreement with the host country at the beginning of the program to determine how exercises will be used. Exercises can be used as a training tool or a formal measure of the success of a program. In most cases, exercises will probably be used for both training purposes as well as to measure the success of the program from both the host country and the U.S. funding perspective. It is important that the parties agree upon the informal and formal exercises, how and who will conduct the exercises, and how the outcomes will be used.

Exercises are about Learning

Informal exercises are most effective when they are used for training purposes without the stigma of a grade attached, and when they are about learning, gap recognition, and means of improving the system. Exercises provide an opportunity to discover performance gaps between individual components or elements of the system through system integration. These exercises offer a new means of determining gaps in individual or program capabilities for the host-country. When exercise observations are made by an outside party, even if it is the funding party, the exercise should be conducted with consideration for the local culture.

Formal exercises will likely be associated with milestones for programmatic funding and host country cooperation to meet the overall goals of the program. As the cooperative programs move toward meeting international regulations or guidelines, third party (i.e., independent) participation in the formal exercises becomes more important.

Preparation for Exercises

On the whole, the IOC assessment effort can be both costly and valuable. DoD implemented a formal plan to measure attainment of the IOC and the FOC within each country for the CBEP. While the overall requirements among the FSU countries were the same, the exact specifications and testing for each country differed based on the government-to-government agreements in the areas of disease surveillance, cooperative research, threat infrastructure elimination and engagement of selected host-country scientists. The FSU countries engaged are good examples of cooperation in meeting the overall goals of the biological CTR program, and they illustrate how each country poses unique challenges in meeting the formal metrics of the IOC.

As noted above, a number of informal table-top and field exercises were conducted in preparation for the formal IOC exercise. These table-top and field exercises were used as training events. The training exercises were conducted in a nonthreatening manner, so that the host country participants could gain experience and confidence in the use of the new infrastructure (e.g., biological safety level-2 laboratories and secure repositories for especially dangerous pathogens), equipment, diagnostic assays and standard operating procedures that were instituted through DTRA funding. The trainers consisted of DoD sponsored in-country prime contractor personnel and subject-matter experts from a number of government agencies that came to the FSU periodically. Preparation for the IOC took approximately one year, once the in-country disease surveillance system was sufficiently complete to be tested. Because DoD had funded all aspects of the infrastructure, equipment, diagnostic assays, and standard procedures, the IOC was truly a test of the complete disease surveillance system.[a]

Each organization that provided infrastructure, training, reagents, supplies, and procedures was involved in training of the local nationals so that they could meet the objectives of the IOC. Each organization was also involved in preparation of table-top exercises, and field exercises provided additional training after each event to rectify gaps. Metrics were determined for each portion of the disease surveillance program, based on standard operating procedures. The procedures were based on DTRA's requirements for biosafety, security, disease recognition and diagnosis, and international and host government regulations, guidelines, and policies. To train and conduct informal and formal exercises with U.S. personnel, all written materials related to biosafety, biosecurity, training materials, equipment manuals, and procedures of the disease surveillance system were translated into English and the host country's language. Because DoD had a strong influence on the decisions regarding infrastructure, supplies, reagents, training, and procedures, it was easy to implement a systems approach for the disease surveillance program. DTRA's approach was to work closely with FSU partners essentially to clone U.S. procedures (i.e., to attain IOC). Those procedures are, of course, most familiar to DoD and if a partner country asks to be trained on the U.S. procedures then using them makes sense; but other procedures might meet the disease identification and reporting goals. It seems that developing metrics that assess outcome and that do not specify a particular process (i.e., a more flexible approach) would work better in countries with different preexisting infrastructure and operations.

The resources required by DoD, the prime integrating contractor, and the supporting USG agencies were significant to meet the IOC. The IOC assessment was not only a test of the disease surveillance system but of DTRA, the prime integrating contractor and the U.S. government agencies involved in the training of local nationals. All groups worked to prepare the FSU countries to pass.

Exercises and Nunn-Lugar Global Cooperation[b]

The IOC and FOC represent major and critical tests of major procurements by DoD and may not be appropriate for countries with small engagement projects, but the concept of testing (exercises) should be similar, if the question is how well a country is safely and securely responding to especially dangerous pathogens.

Exercises under an expanded CTR Program could be similar in concept to what was previously tested in the FSU, but it is likely that DTRA will only be funding discrete parts of the disease surveillance system or supplementing existing programs. This will require harder to develop exercises because the procedures, diagnostic kits, safety and security infrastructure, and controls may be different than those funded by DoD. DoD's approach has been to work closely with FSU partners essentially to clone U.S. procedures (i.e., to attain the IOC as noted above). If DoD is only supplementing the partner countries disease surveillance system, the exercises will need to be more flexible in training/testing general concepts and procedures, rather than prescribed diagnostic assay, procedures, and infrastructure.

DoD had significant influence in countries like Azerbaijan, Georgia, Kazakhstan, and Uzbekistan due to the funding supplied for the host country's infrastructure and systems for acquiring, testing, securing, and reporting results of suspected especially dangerous microorganisms. As the economic situation has improved in some countries, DoD's influence has decreased. Where the economic situation has improved, some countries have decided to move forward with their own initiatives without DoD's funding, and replacing equipment that DoD was using to standardize diagnostic platform across countries. As a result, while the overall objectives of the CTR-implemented program disease surveillance system may remain the same, DoD will need to be more flexible in determining how to measure the outcomes. In other countries, the political situation is such that DoD may have less access or less timely access to facilities where DTRA has funded infrastructure, training, supplies, and procedures. Compare, for example, the free access that the U.S. government enjoyed when establishing the program in Georgia versus the efforts in Uzbekistan. Where it is more difficult to determine the overall success of the program due to a lack of access to facilities that DoD may have built, supplied, and trained, the international regulations or guidelines may still be achieved. DoD is also reexamining the infrastructure, equipment and supplies based on what can be sustained, but it is not clear how metrics will be applied to those countries for which that goal is not currently attainable as a result of economic unsustainability.

[a] Host country salaries were not paid by DTRA.

[b] The NRC Committee on Global Security Engagement has called this global cooperation CTR 2.0 (NRC, 2010).

Time and Change

> **DoD's metrics and planning process should factor in more explicitly both planned and unplanned change over time. During the phases of active DoD involvement in a CTR project and afterward during sustainment, which is its own stage requiring resources (budgets, equipment, and trained people), clearer planning for how changes and metrics results will feed into decision making will make the metrics more credible and useful for both DoD and the partner country.**

As mentioned in Chapter 2, the DoD Metrics Report deliberately does not consider future missions or changes in objectives,[31] and although it notes that programs may change and expand to new geographic areas, some sections of the DoD Metrics Report make CTR appear to be a static program. Given that the circumstances in Africa, South East Asia, India, and elsewhere are very different from the contexts in which DoD has worked in the past, the metrics used will have to adjust accordingly, particularly those regarding capacity building and other more-difficult-to-measure objectives. Also, should the objectives (and therefore the metrics) remain unchanged if the budget available doubles? If it divides in half? If the goal is to modernize a country's capabilities, how does DoD think about obsolescence of the equipment and techniques (the capability) it provides? DoD says that it will adjust and adapt its metrics by revising the Report annually, but without a vision for how change fits into the Program, it will be difficult for DoD to adapt well.

CTR projects change through planned evolution as they progress through different phases and timelines, scope, and even objectives change in unplanned ways as the environment and resources change. Indeed, changes will likely be made in response to results as reported through the metrics. Because of this iterative linkage, having more detail on the objectives development process and on how the metrics will be used in that process would be helpful in defining useful metrics going forward.

For example, consider the Central Reference Laboratory (CRL) planned for Ukraine. DoD committed to paying for the design and construction of a new CRL facility for work on especially dangerous human pathogens. In the last few years, DoD became concerned that the facility would cost too much, both in capital costs and in maintenance and operation costs, which would undermine the independent sustainability of the project. To follow a more sustainable path, DoD proposed to the Ukrainian partners to use the interim CRL (an upgraded existing facility) rather than create a new green-field site. Ukraine reluctantly agreed. This adaptation reflected a better understanding of budget realities in both countries and experience building a new CRL elsewhere. This change is the kind of adaptation and course correction DoD wants to see in its programs. However, if DoD does not revise its metrics, the metrics will not reflect this positive outcome: A metric for progress on construction of the new CRL building would show no progress, but the capability might be in place more quickly and sustainably using the alternative approach.

---

[31] "The metrics described in this report are designed to measure appropriately the impact of each CTR program area, as the CTR program is currently constituted for Fiscal Year 2010, and does not attempt to speculate on what metrics might be appropriate for future areas of programmatic or geographic expansion. Neither do the metrics attempt to determine whether the activities of the CTR program are the "right" activities. The metrics described in this report are intended to best measure the effectiveness of the CTR Program in conducting these established programs. These metrics are not intended to revise the method for establishing these objectives." (DoD 2010).

There is a school of thought that says that metrics should not change over the course of a project, even if other factors do. The committee notes that as social scientists have assisted the USAID in trying to develop rigorous metrics for its programs, an approach has been recommended that is, in essence, a set of experiments that attempt to establish causal links between aid and outcomes. Communities are identified either as recipients of a particular kind of aid project or as a control group. The projects must proceed with the same metrics and the same aid established at the outset to ensure the integrity of the experiment. However, controlled experiments are not the only way to measure outcomes and do not appear to be well suited to CTR programs. The needs and objectives of each project may be unique to the partner country and the environmental factors and objectives for the program change. It may be trivial to establish causal links to some desired outcomes and simply impossible for others.

If sustainability is an important objective for the CTR Program, change needs a more prominent role in DoD's thinking, because it will be difficult to address sustainability without considering time and change explicitly. In the DoD Metrics Report, sustainability is mentioned as an objective for some programs, but the only factor considered in evaluating the sustainability of a project (the only metric) is cost (i.e., can the partner country pay for the program for the few years after the end of the CTR project). There is more to sustainability than cost and while measuring sustainability may be difficult, it is not impossible.

Evolution of Metrics Over Program Lifecycle

There is a natural evolution of cooperative threat reduction programs that should in principle be reflected in the metrics that measure impact and effectiveness. The metrics that are most useful and informative in the early stages of a program, where defining the ultimate program scope and priorities are a central focus, will generally differ from those that warrant closest attention toward the end of a program's life cycle, where issues of sustainment and close-out come to the forefront. Some aspects of this evolution are already evident at least implicitly in the metrics described in the DoD Metrics Report, for example, the increased emphasis on sustainability and on partner country follow-through. The committee believes this concept of evolution of metrics should be more explicitly considered in future iterations of CTR metrics so that the progression of CTR activities through the successive phases would be more visible and prominent.

In many fields of application, such a progression is described in terms of "maturity models." Examples include:

- ***Quality management***. in his book *Quality is Free*, Crosby (1992), introduced a five-stage quality maturity model, which described the phases of a quality improvement program from inception to maturity, with the nature of the challenges evolving along the way. Although some metrics apply across all stages, e.g., measures of quality, others come and go as a program progresses through the five stages.
- ***Information technology, enterprise architecture, and software***. Maturity models have been developed for many facets of information technology and a vast literature exists that could, with appropriate adaptations, be instructive for CTR programs. For example, for building enterprise architectures (EA, somewhat analogous to building threat reduction architectures) five stages have been defined: Stage 1: Creating EA Awareness; Stage 2: Building the EA Management Foundation; Stage 3: Developing the EA; Stage 4:

Completing the EA; Stage 5: Leveraging the EA to Manage Change. Clearly, some adaptation to the threat reduction context is necessary, but the basic stages—awareness, planning and design, building, and so forth, have parallels. For an entry point to the literature on enterprise architecture maturity models, see for example GAO (2003), and the references therein. DoD and other agencies have their own agency-specific products in this area as well, e.g., Department of Defense Architecture Framework.

Many maturity models have their roots, directly or indirectly, in concepts of process control or quality control pioneered by Shewhart, Deming, and others in the first half of the twentieth century which defined process and quality improvement in terms of an iterative improvement cycle, often described as plan, do, check, act, or PDCA. Depending on the field of application, the basic PDCA, stages may also be described as planning, implementation, evaluation, specification, production, inspection, always followed by iterative improvement cycles repeating the same basic steps.

The committee is not recommending a particular maturity model or vocabulary because the concepts are most useful if adapted to the problem at hand. The key point is that for CTR programs there is often a natural progression of stages from conception to completion, and natural metrics for each stage. Although details may differ from program to program, a series of questions often arise at each stage:

- What is the scope of the threat reduction challenge? Metrics can include the numbers and types of objects (sites, facilities, weapons, materials, infrastructure, people) that collectively make up the threat to be reduced.
- What tools are needed to manage or reduce the threat? Metrics include the delivery of specific tools (equipment, training, procedures, personnel, etc.).
- Are the tools being used, and used properly? Tests, exercises and routine evaluations can furnish metrics on the extent to which the tools are being used.
- Are the threat reduction capabilities, as implemented, effective? Having and using the tools is not by itself sufficient to ensure effectiveness. Metrics that measure results rather than merely outputs are needed at this stage.
- Is there a system in place to maintain and ensure effectiveness on an ongoing basis, and initiate corrective actions if necessary? Many of the sustainability metrics that are now being introduced are aimed at this stage.

At any given time, a program is likely to be focused mainly on one or two stages. Nevertheless, the committee believes it is helpful to consider and present the larger context as well. The overall scope of the threat reduction challenge and the portion that a given program is attempting to tackle is important contextual information that helps keep the overall objectives in focus over the overall program lifecycle even as more specific output- and activity-based metrics are the central focus during a given phase of a program. Even if different metrics are *applied* at various stages during the conduct of a project, it would be useful to have *developed* the various metrics from the start, recognizing that they might have to change.

OTHER MAJOR ISSUES FOR CTR IN THE FUTURE

Scope of the CTR Mission

It is beyond the scope of this study to say what is in or outside of the CTR mission. Some authorizers and appropriators questioned whether CBEP is in fact a defense mission. Fundamentally, whether DoD should be the agency to carry out the CBEP mission is only secondarily a metrics question. The primary question is whether the U.S. Government wishes to prioritize the work done under CBEP and what mix of government agencies is best equipped to carry it out. The increases in budget and scope to date, as well as the National Strategy for Countering Biological Threats, indicate that CBEP is a growing priority to the Administration, and DoD's involvement reflects a conscious choice to use DoD because of its experience establishing on-going cooperative medical and bioresearch ventures (e.g., NAMRU-3 in Cairo and NAMRU-6 in Lima), conducting biodefense research, and working on threat reduction. Critics may dispute the decision to support the mission, to give DoD responsibility for the mission, or to give Defense Threat Reduction Agency responsibility for implementation, and they can legitimately point out that difficulties in developing reliable direct metrics, for the program's impact and effectiveness raise the programmatic risks of the program, but mixing these issues with questions about the metrics themselves confuses matters and makes it more difficult to make progress in the program and in the debate about the program.

Recognizing Success/Completion

Finally, defining and measuring completion—how do we know when we are done?—and sustainability—will the changes take hold and will the partner nation support and sustain the programs when U.S. funding stops?—are critically important for the CTR programs, particularly the capacity building programs. What these mean and how they should be implemented and measured for a given program should be part of the formulation of objectives. There is a mismatch in the vision of sustainability and measuring completion of the program among different CTR decision makers in the Administration and on Capitol Hill. One vision might be called a project view, in which DoD partners with a nation, engages in a set of concrete activities with a well defined beginning and end, and then DoD exits and monitors sustainment after project completion. The other main vision might be called a relationship view, in which DoD partners with a nation, works with the partner to build a joint or multilateral network that is exercised regularly to maintain an on-going relationship with no defined end date. These visions appear mutually exclusive, but there are different phases to capacity building programs: the initial phase may involve intensive efforts and capital expenditures. There should be schedules and milestones for completion of this phase. The long-term relationship that follows may be open ended, but it also should require far less funding, which should allay concerns about programs with no exit strategy.

REFERENCES

Ball, Deborah Yarsike and Theodore P. Gerber. 2004. Will Russian Scientists Go Rogue? A Survey on the Threat and the Impact of Western Assistance. *PONARS Policy Memo* 357. Available at: http://www.gwu.edu/~ieresgwu/assets/docs/ponars/pm_0357.pdf; accessed July 19, 2011.

Davis, Paul C. 2002. Analytic Architecture Capabilities Based Planning, Mission-System Analysis, and Transformation. Santa Monica, CA: Rand Publishing.

Department of Defense (DoD). 2003a. Defense Security Cooperation Agency. Directive Number 5105.65 issued October 31, 2000 (Incorporating Change 1, September 23, 2003). Available at: http://www.dtic.mil/whs/directives/corres/pdf/510565p.pdf; accessed January 11, 2012.

DoD. 2003b. Defense Acquisition System. Directive Number 5000.1 (DoD 5000.1), May 12, 2003. Available at: http://www.ntip.navy.mil/cap/documents/500001p.pdf; accessed January 11, 2012.

DoD. 2003c. Operation of the Defense Acquisition System. Instruction Number 5000.2 (DoD I 5000.1), May 12, 2003. Available at: http://www.carlisle.army.mil/dime/documents/JPLD_AY08_Lsn%207_Reading%203_DoDI%205000-2.pdf; accessed January 11, 2012.

DoD. 2011. Defense Acquisition Guidebook. Available at: http://at.dod.mil/docs/DefenseAcquisitionGuidebook.pdf; accessed January 11, 2012.

Government Accountability Office (GAO). 2003. Information Technology, A Framework for Assessing and Improving Enterprise Architecture Management (Version 1.1). GAO-03-584-G. Available at: http://www.gao.gov/new.items/d03584g.pdf; accessed January 11, 2012.

GAO. 2011. Border Security. Preliminary Observations on Border Control Measures for the Southwest Border. GAO-11-374T. Available at: http://www.gao.gov/assets/130/125500.pdf; accessed January 11, 2012.

International Science and Technology Center (ISTC). 2011. Statement of the 53rd Governing Board of the ISTC. 28.06.2011. Available at: http://www.istc.ru/istc/istc.nsf/va_WebPages/GBStatement53EngPrint; accessed July 19, 2011.

Keeney, R.L. and Raiffa H. 1993. Decisions with Multiple Objectives: Preferences and Value Tradeoffs. Cambridge: Cambridge University Press.

Kirkwood, C. W. 1997. Strategic Decision Making: Multiobjective Decision Analysis with Spreadsheets. Belmont, CA: Brooks/Cole, Cengage Learning.

Nacht, M. 2009. Memorandum for Assistant to the Secretary of Defense for Nuclear and Chemical and Biological Defense Programs. Biological Threat Reduction Program Strategic Planning Guidance and Program Objectives. Available through the Public Access File for this study.

National Research Council (NRC). 1997. Controlling Dangerous Pathogens: A Blueprint for U.S.-Russia Cooperation. Washington, D.C.: National Academies Press.

NRC. 1997. Proliferation Concerns: Assessing U.S. Efforts to Help Contain Nuclear and Other Dangerous Materials and Technologies in the Former Soviet Union. Washington, D.C.: National Academies Press.

NRC. 1999. Protecting Nuclear Weapons Materials in Russia. Washington, D.C.: National Academies Press.

NRC. 2006a. Biological Science and Biotechnology in Russia: Controlling Diseases and Enhancing Security. Washington, D.C.: National Academies Press.

NRC. 2006b. Strengthening Long-Term Nuclear Security: Protecting Weapon-Usable Material in Russia. Washington, D.C.: National Academies Press.

NRC. 2006c. The Fundamental Role of Science and Technology in International Development: An Imperative for the U.S. Agency for International Development. Washington, D.C.: National Academies Press.

NRC. 2006d. Threat Agent Detection Response System Data Base. Washington, D.C.: National Academies Press.

NRC. 2007a. U.S.-Russian Collaboration in Combating Radiological Terrorism. Washington, D.C.: National Academies Press.

NRC. 2007b. The Biological Threat Reduction Program of the Department of Defense: From Foreign Assistance to Sustainable Partnerships. Washington, D.C.: National Academies Press.

NRC. 2008. Improving Democracy Assistance: Building Knowledge Through Evaluation and Research.

NRC. 2009. Countering Biological Threats: Challenges for the Department of Defense's Nonproliferation Program Beyond the Former Soviet Union. Washington, D.C.: National Academies Press.

National Security Council. 2009. National Strategy for Countering Biological Threats. Washington, D.C.

Parnell, G. S., Driscoll, P. J., and Henderson D. L., Editors, 2011. Decision Making for Systems Engineering and Management, 2nd Edition. Hoboken, NJ: Wiley & Sons.

Revill, J. 2009. Developing Metrics and Measures for Dual-Use Education. Research Report for the Wellcome Trust Project on 'Building a Sustainable Capacity in Dual-use Bioethics.' Available at: http://www.brad.ac.uk/bioethics/media/SSIS/Bioethics/docs/Education_metrics_and_measures_2nd_DRAFT.pdf; accessed January 11, 2012.

Weber, A. 2010. Memorandum for Director, Defense Threat Reduction Agency. Subject: Implementation Guidance for: I. Cooperative Biological Engagement Program (CBEP) Activities in Pakistan; II. Improving Implementation Flexibility Across CBEP; and III. Pursuing Implementing Arrangements with International Organizations Relevant to the CBEP Mission.

GLOSSARY AND LIST OF ACRONYMS

Activity-based metrics: Measures of performance or progress focused on specific aspects of an individual activity or project.

Adversary measures: Measures that assess the potential or actual actions of adversaries that can have a direct impact on the process or system.

Aspirational goal: A desired performance level or milestone beyond the minimum requirement. Such goals are useful for guiding the program or project toward the larger objective and can be the basis for performance incentives.

Attribute: A characteristic of a system or process.

Benefit cost measure: A metric reflecting benefits attained relative to the costs, which may be monetary or other costs.

BS&S: Biological safety and security

BTRP: Biological Threat Reduction Program

BW: Biological weapons

Capabilities based planning: Capability based planning focuses on the planning, engineering, and delivery of strategic capabilities by defining objectives, identifying what capabilities are needed to achieve the objectives, and then determining how to ensure that those capabilities are developed operating in an uncertain real-world environment.

CBEP: Cooperative Biological Engagement Program

CBP: Capabilities Based Planning. In context, CBP may also mean Customs and Border Protection.

COCOM: Combatant Command

Constructed metric: A metric that is developed by relying on integral judgments of experts.

Constructed scale: The range of a metric that integrates several performance attributes into one artificial scale (e.g., a 5-star safety rating).

Cost effectiveness: Cost effectiveness assesses attainment of an objective or task relative to the costs.

Criteria: Factors that differentiate alternatives.

CRL: Central Reference Laboratory

CTR: Cooperative Threat Reduction

CWEP: Chemical Weapons Elimination Program

Direct measure: Measurement of an attribute that itself embodies the status of a threat to be reduced. Change in a direct measure indicates impact.

Direct outcome measure: A measure based on attributes that are the direct results of an activity or program.

DoD: Department of Defense

DSCA: Defense Security Cooperation Agency

DTRA: Defense Threat Reduction Agency

EDP: Especially dangerous pathogen

Efficiency measure: A measure of how resources are used in a project or program.

Environmental measure: A factor that could have a direct or indirect impact on the system or process but is not under the control of the program or project managers.

FOC: Full Operating Capacity

FSU: Former Soviet Union

Higher level program performance metrics: These metrics are developed to measure the impact and effectiveness of a program or project on a more strategic level, rather than on an activity level

ICBM: Intercontinental ballistic missile

Impact and effectiveness: The extent to which the program or project is reaching strategic objectives and is achieving results.

Inputs: The resources invested into a project or program, or the activities undertaken that are meant to lead to outcomes.

Input measure: A measure to identify the resources or activities provided for a project or program.

IOC: Initial Operating Capacity

Lagging indicator: A measure that is available after the current activity allowing corrective action only for future processes and operations.

Leading indicators: Measures that provide early indicators of impact and effectiveness and allow managers to take corrective action on the process if required.

Maturity model: A tool to assess progress throughout the duration of the project or program, moving from stage to stage of the project or program.

Metric: A standard or indicator of measurement.

Minimum performance level: The minimum level at which a program or project can perform to be considered within acceptable performance range.

Natural scales: An already available means of measuring the impact and effectiveness of a project/program against the stated objectives.

NWSSP: Nuclear Weapons Safety and Security Program

Output: A direct, measurable result of a process, activity, or project that may or may not reflect impact.

Outcome: An overall result of an activity, project, or program.

Performance measure: Describes how well a subsystem, system, or process meets its required performance.

Program: An organized effort toward an overall goal and a set of objectives supporting that goal. Typically, a program comprises a set of projects or activities carried out under the program.

Program goals: Goals designated for the overall program as opposed to the specific goals of the projects that are implemented to carry out the program.

Program management measure: A measure that is useful for measuring the cost-effectiveness and efficiency of the project/program, but not necessarily the impact and effectiveness of the program toward the overall program objectives.

Program objective: An objective that is linked to threat reduction and is applicable to the program as a whole rather than to the parts of a program.

Project: A set of activities carried out to support one or more program objectives. Projects usually constitute the actual implementation of a program.

Project goals: Goals developed to guide a discrete project as opposed to an overall program.

Project metric: Metrics designed to measure the impact and effectiveness of the project against the project goals.

Proxy measure: An indirect measure of the impact and effectiveness of an aspect(s) of a program or project.

Resource measure: A measure that identifies the resources (e.g., dollars, people's time, or materials) used by a system or process.

SOAE: Strategic Offensive Arms Elimination Program

Strategic metrics: A metric developed to determine the impact and effectiveness of a program or project against the strategic goals.

Sustainability: The ability and likelihood of a project or program being maintained into the future.

Sustainment: The continuation of the efforts, impacts, and outcomes of a program or project into the future, eventually solely by the partner country.

Swing-weight analysis: A swing weight matrix defines the importance and range of variation for a set of metrics. In the analysis, a metric that is very important to the decision is weighted higher than a metric that is less important. A metric that differentiates between alternatives is weighted more than a metric that does not differentiate between alternatives.

Threat measure: A measure that assesses capabilities and the intent of a potential opponent.

WMD-PPP: Weapons of Mass Destruction Proliferation Prevention Program

APPENDIX A

CONGRESSIONAL MANDATE

Public Law No: 111-84. National Defense Authorization Act for Fiscal Year 2010

SEC. 1304. METRICS FOR THE COOPERATIVE THREAT REDUCTION PROGRAM.

(a) Metrics Required- The Secretary of Defense shall develop and implement metrics to measure the impact and effectiveness of activities of the Cooperative Threat Reduction Program of the Department of Defense to address threats arising from the proliferation of chemical, nuclear, and biological weapons and weapons-related materials, technologies, and expertise.

(b) Secretary of Defense Report on Metrics- Not later than 270 days after the date of the enactment of this Act, the Secretary of Defense shall submit to the appropriate congressional committees a report describing the metrics developed and implemented under subsection (a).

(c) National Academy of Sciences Assessment and Report on Metrics-

(1) ASSESSMENT- Not later than 30 days after the date on which the report is submitted by the Secretary of Defense under subsection (b), the Secretary shall enter into an arrangement with the National Academy of Sciences under which the Academy shall carry out an assessment to review the metrics developed and implemented under subsection (a) and identify possible additional or alternative metrics, if necessary.

(2) REPORT- The National Academy of Sciences shall submit to the appropriate congressional committees and the Secretary of Defense a report on the results of the assessment carried out under paragraph (1).

(3) SECRETARY OF DEFENSE REPORT-

(A) Not later than 90 days after receipt of the report required by paragraph (2), the Secretary shall submit to the appropriate congressional committees a report on the assessment carried out by the National Academy of Sciences.

(B) The report under subparagraph (A) shall include the following:

(i) A summary of the results of the assessment carried out under paragraph (1).

(ii) An evaluation by the Secretary of the assessment.

(iii) A statement of the actions, if any, to be undertaken by the Secretary to implement any recommendations in the assessment.

(C) The report under subparagraph (A) shall be submitted in unclassified form, but may include a classified annex.

(d) Funding- Of the amounts appropriated pursuant to the authorization of appropriations in section 301(20) or otherwise made available for Cooperative Threat Reduction Programs for fiscal year 2010, not more than $1,000,000 may be obligated or expended to carry out paragraphs (1) and (2) of subsection (c).

(e) Appropriate Congressional Committees Defined- In this section, the term 'appropriate congressional committees' means--

(1) the Committee on Armed Services, the Committee on Appropriations, and the Committee on Foreign Affairs of the House of Representatives; and

(2) the Committee on Armed Services, the Committee on Appropriations, and the Committee on Foreign Relations of the Senate.

APPENDIX B

REPRODUCTION OF THE DEPARTMENT OF DEFENSE COOPERATIVE THREAT REDUCTION METRICS REPORT

Delivered by Department of Defense to Congress

September 2010

Introduction

This report is submitted in accordance with Section 1304 of the National Defense Authorization Act for Fiscal Year 2010, Title XIII of Public Law 111-84. Section 1304 provides that the Secretary of Defense shall develop and implement metrics to measure the impact and effectiveness of activities of the Department of Defense (DoD) Nunn-Lugar Cooperative Threat Reduction (CTR) Program. The metrics to be developed would measure the effectiveness of projects and activities undertaken by DoD to prevent the proliferation of chemical, nuclear, and biological weapons and weapons of mass destruction (WMD)-related materials, technologies, and expertise.

Background

In the National Academy of Sciences' (NAS) 2009 report, *Global Security Engagement: A New Model for Cooperative Threat Reduction*, NAS affirmed the usefulness of the traditional metrics of the DoD CTR Program for evaluating effectiveness in traditional CTR WMD elimination or site security program areas, in which concrete actions such as chemical munitions, warhead delivery systems, and launchers destroyed or chemical or nuclear warheads and weapons materials secured clearly lead to threat reduction. Such activities break down into discrete events with clear end points, and thus metrics that can be incrementally reported, as on the CTR Scorecard (Appendix 1), continue to be appropriate for those types of programs. However, the NAS report accurately noted that CTR program responses to new, often less quantifiable threats in this century will require new methods to measure success. The challenge is to find measureable performance indicators that capture the true value of program activities that may not lend themselves readily to raw quantitative measurement, thus making it more difficult to document the relative success of the activity.

In recent years, the CTR program has increasingly become involved in capacity-building efforts, notably in the Cooperative Biological Engagement Program (CBEP), formerly the Biological Threat Reduction Program (BTRP), and the WMD Proliferation Prevention Program (WMD-PPP). In these programs, threat reduction is accomplished through building the capacity of the partner state to conduct surveillance to deter or detect and report or respond to emerging biological and WMD trafficking threats. Similarly, the CTR Program is building the capacity of the Russian Federation to sustain upgraded physical protection systems at nuclear weapons storage sites. Although increased partner state capacity makes an important contribution to a reduction in the WMD threat to the United States, this is more difficult to quantify than numbers of missiles or other forms of strategic offensive arms destroyed as in the traditional programs.

CTR's prevention mission is difficult to quantify. Although a number of output measures can be counted, they are proxies for the outcome of deterring terrorists and proliferators from accessing WMD and related materials and expertise. We can measure the amount of equipment provided and the number of training events conducted or scientists engaged; however, we need better measures to show that these efforts actually result in changed practices or additional effectiveness.

As an example that makes evident the findings in the NAS report and reflects a matter DoD has recognized in connection with the CTR Program, a metric based on the number of Zonal Diagnostic Laboratories built under the BTRP may not accurately reflect the partner country's enhanced capability for pathogen detection and reporting. Indeed, such a metric could create unintended incentives to continue building additional laboratories that do not make a positive contribution to threat reduction, when what is really needed might be additional training to make better use of the existing laboratories' capabilities for pathogen detection. In WMD-PPP, the raw number of radars, sensors, and patrol boats provided for border security, although useful, may not be as important as the effectiveness of the information sharing among those systems, which is a function of system integration, training and effective decision making rather than of the number of systems provided.

In light of these issues, Congress has mandated the development and implementation of metrics to measure the impact and effectiveness of projects and activities of the CTR program to address threats arising from the proliferation of chemical, nuclear, and biological weapons and WMD-related materials, technologies, and expertise. Additionally, Section 1304 requires submission, no later than 270 days after the enactment of the Act, of a report describing the metrics developed and implemented. This report addresses these Section 1304 requirements.

Discussion

The metrics described in this report are designed to measure appropriately the impact of each CTR program area, as the CTR program is currently constituted for Fiscal Year 2010, and does not attempt to speculate on what metrics might be appropriate for future areas of programmatic or geographic expansion. Neither do the metrics attempt to determine whether the activities of the CTR program are the "right" activities. The metrics described in this report are intended to best measure the effectiveness of the CTR Program in conducting these established programs. These metrics are not intended to revise the method for establishing these objectives.

Figure 1 illustrates the organization of the Department of Defense as it relates to the CTR program. CTR executes programs as directed in program guidance from the Deputy Assistant to the Secretary of Defense (Treaties and Threat Reduction) (DATSD(T&TR)). This program guidance provides instruction on how to implements policy guidance issued by the Office of the Under Secretary of Defense for Policy (USD/P), which has been coordinated with all relevant agencies within the Executive Branch.

Those responsible for each CTR program area, in concert with applicable stakeholders, reviewed their specific program objectives and either established new program metrics or confirmed the validity of existing program metrics. Although not culled out specifically in this report, except as noted in the CTR Scorecard (Appendix 1), the metrics currently in use for the Strategic Offensive Arms Elimination (SOAE) program will not change. The results of this review and analysis process are provided in this report. Each report section provides a brief program description, methodology for metrics development, and a description of metrics that includes reporting and re-evaluation activities.

Conclusion

CTR program metrics serve as a means to evaluate how well program objectives have been accomplished. Several traditional CTR program metrics are still very applicable and useful measures of CTR Program threat reduction achievements. For those programs involved in capacity-building efforts, notably the CBEP, basic metrics need to be specifically tailored and evaluated individually at a project level. A number of program benefits that would be useful and meaningful in the larger USG threat reduction effort and are routinely considered in establishing execution plans, such as partner country contributions and commitments, or ability to leverage multiple sources of expertise or funding, were considered but not recommended since these benefits ultimately do not trace back directly to the CTR program objectives. As the CTR program continues to evolve, we will continue to assess its metrics to ensure they best represent achievement of new objectives.

Figure 1. The CTR program within the DoD organizational structure

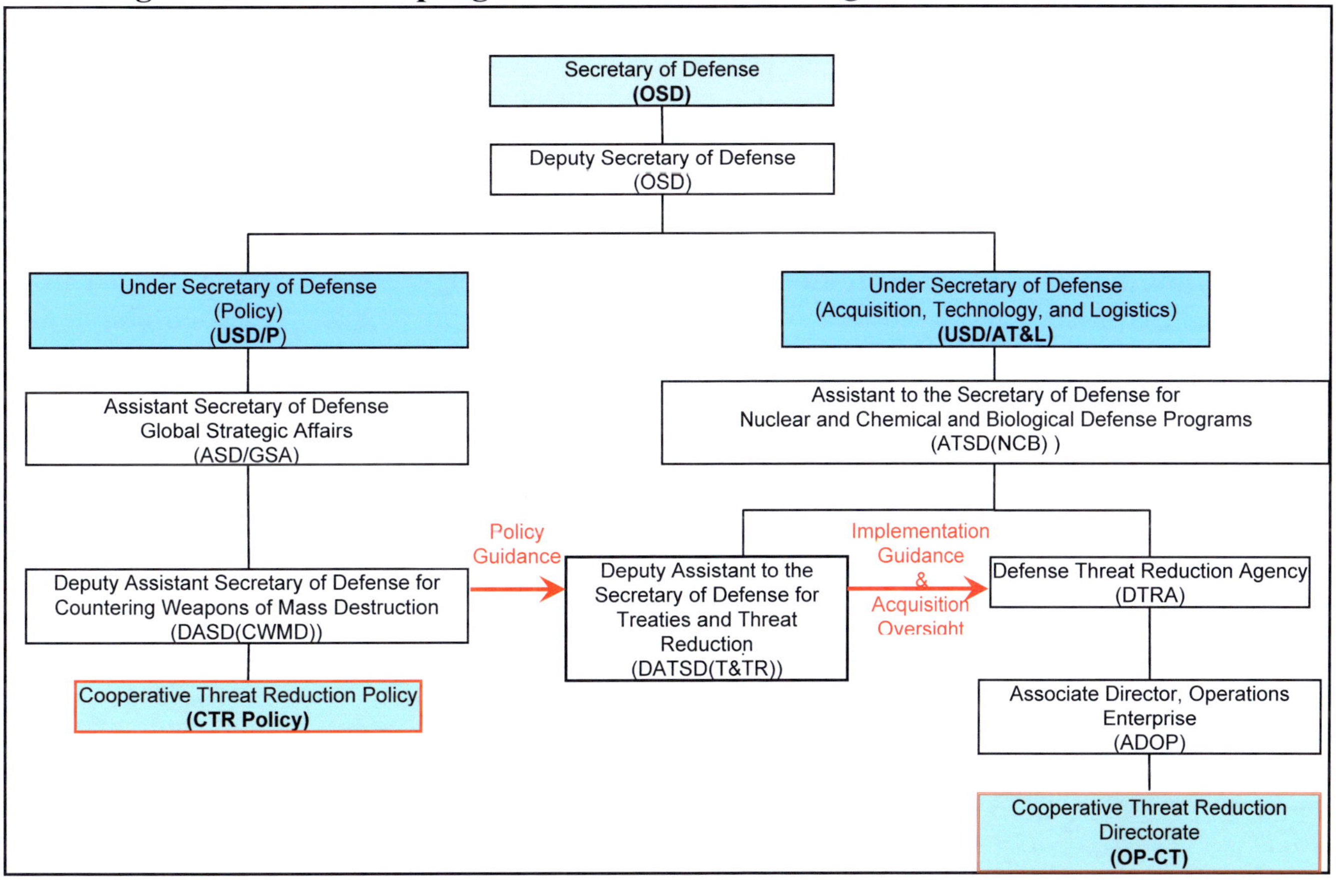

Cooperative Biological Engagement Program

Program Description

The Cooperative Biological Engagement Program (CBEP) has four main objectives that are identified in the current policy and implementation guidance issued to the Defense Threat Reduction Agency (DTRA) for implementation:

1. Secure and consolidate collections of especially dangerous pathogens (EDPs) and their associated research at a minimum number of secure health and agricultural laboratories or related facilities;
2. Enhance partner country/region's capability to prevent the sale, theft, diversion, or accidental release of biological weapons (BW)-related materials, technology, and expertise by improving biological safety and security (BS&S) standards and procedures;
3. Enhance partner country/region's capability to detect, diagnose, and report endemic and epidemic, man-made or natural EDPs, bio-terror attacks, and potential pandemics; and
4. Ensure the developed capabilities are designed to be sustainable within each partner country/region's current operating budget.

The fourth objective identified above, sustainability, is a cross-cutting objective to ensure that the capabilities provided by the CBEP are adapted and owned by the partner country. The ultimate aim for CBEP implementation is that the partner country take ownership of a sustainable capacity to conduct biosurveillance, meet BS&S standards, and conduct research effectively on its own. Promoting partner country ownership is one of the critical program objectives and one of the most significant CBEP challenges.

The CBEP's supplemental objectives – objectives noted as supplemental in the guidance received – include facilitating the engagement of partner country's/regional scientific and technical personnel in research areas of interest to both the partner country/region and the United States and eliminating any BW infrastructure and technologies encountered in a partner country/region. CBEP assistance also supports host country objectives in meeting the World Health Organization (WHO) International Health Regulations (IHR).

CBEP implementation to meet these objectives includes collaborative biological research (CBR), biosurveillance, and enhancement of BS&S. CBR efforts bring together U.S. and international scientists in cooperative research of mutual interest in support of threat reduction priorities. Biosurveillance projects include providing training to animal and human health experts on disease recognition, diagnosis, and reporting; support for laboratory diagnostic equipment and reagents; and implementation of the Electronic Integrated Disease Surveillance System (EIDSS). BS&S enhancements include bioethics training, training on safe clinical and laboratory practices; installation of security systems at laboratories; implementation of the Pathogen Asset Control System (PACS) to track inventory; and support for development of biorisk management processes and procedures.

Methodology for Development of New Metrics

Developing metrics for the CBEP presented several challenges. First, the program seeks to achieve concrete improvements in disease detection, biological safety, and research systems in an environment not fully under its control, with the result that accountability and analysis of program impact are complex, difficult matters. Also, the metrics for the CBEP need to establish a solid, traceable link from CBEP-funded efforts to concrete threat reduction achievements.

To address these challenges and establish actionable metrics, the CBEP adapted a Joint Capabilities Integration and Development System-like process to develop mission and performance measures. This effort took place over the course of one year of collaborative development. The CBEP metrics team first developed an initial capabilities document that lays out the key capabilities required in the program's strategic guidance and the attributes supporting these capabilities. From these attributes, the team then developed measures of effectiveness (MOEs) to assess mission accomplishment. These MOEs are directly traceable to the program's policy and implementation guidance. Finally, the team developed measurement values to serve as indicators that the critical MOEs have been achieved. The objective and minimum thresholds for these measurement values indicate the desired end state and the minimally acceptable level of achievement respectively.

Throughout the process, the CBEP engaged program stakeholders from across disciplines as diverse as clinical practice, epidemiology, laboratory, biosafety, veterinary, biosecurity, systems engineering, and training. The use of small working groups by discipline and progressive reviews by outside parties and program leadership enhanced the accuracy, comprehensiveness, and usefulness of the final product.

Description of New Metrics

The table in Appendix 2 depicts the linkages and associations across 6 program objectives, 5 program capabilities, 10 attributes, 20 MOEs, measurement values, and minimum/objective values.

The CBEP will utilize these revised metrics throughout planning, implementation, and final assessment of country/region engagements.

Using a systems engineering approach, all project officers will use metrics to plan and execute work, including designing a concept of operations to support MOEs in a partner country or region. The project implementation team will focus on achieving a capability in a partner country or region as defined by the MOEs.

Interim performance against the metrics will be assessed through a seamless, collaborative test and evaluation process that will include tabletop and field training exercises, as well as informal assessments by CBEP collaborators. In addition, ongoing proficiency and competency testing will assess capability levels, target areas for improvement, and build capacity on the micro level. Results will also be fed back to the implementation teams, allowing them to make adjustments to their execution plans. Feedback from CBEP collaborators is a part of each engagement activity in each country or region, and the program holds an integrated country review at least every quarter of the year to identify issues, necessary changes in approach, action items, and those responsible for taking action. This iterative, ongoing assessment approach helps to ensure that enhancements are taking place and that the implementation approach is delivering results.

In conducting formal, large-scale assessments of program achievements relative to the metrics, the CBEP will leverage international authorities such as the WHO, the Food and Agriculture Organization of the United Nations (FAO), or the World Organization for Animal Health (OIE) to seek third-party validation of program accomplishments. These formal reviews will assess the cumulative results of CBEP engagement and identify areas for continuing cooperation.

All of the above assessment efforts will document performance relative to the 20 MOEs identified in Appendix 2. The use of multiple metrics allows for a thorough, broad evaluation of program performance; however, it makes providing a succinct, top-level report of performance a challenge. The CBEP plans to report the cumulative performance of each country against the comprehensive metrics on an annual basis. This summary report would offer an overview of each country's "proficiency" relative to the CBEP objectives. Proficiency could be indicated in some manner on a summary graphic; however, the complexity of the program does not lend itself to simplification into a single scorecard metric.

Chemical Weapons Elimination

Program description

The original program task for the Chemical Weapons Elimination program was to assist the Russian Federation in establishing a chemical weapons destruction facility (CWDF) near Shchuch'ye, located in the Kurgan District, to destroy nerve agent-filled munitions located in the Planovy CW storage facility in a safe, secure, and environmentally sound manner. The assistance provided included the following:

- Site clearing, water drainage, and preparation of the land for construction;
- Evaluating, optimizing, and scaling-up of the Russian-developed nerve agent destruction process;
- Developing, designing, fabricating, and testing of the munitions processing equipment;
- Integrating and systemizing the processing and associated support equipment and facilities; and
- Commissioning the facilities as an acceptable industrial complex under Russian Federation standards and laws.

The United States and other members of the Group of Eight Global Partnership against the Spread of Weapons and Materials of Mass Destruction assisted with completion of this facility.

As construction of the first destruction building and other parts of the complex neared completion, the Russian Federation requested continued U.S. technical assistance to monitor and evaluate the agent destruction system and support equipment maintenance through the remaining agent elimination efforts as a means of assuring continued full functionality of the systems and equipment. The Russian Federation initiated destruction in the first building in March 2009. The second destruction building is still under construction and is scheduled to be complete and in operation in 2011.

Program-Level Metrics

Methodology for Development of New Program-Level Metrics

Technical assistance will primarily take the form of engineering advice to help resolve process and equipment failures, identify potential future failures, and ensure replacement spares are on hand. Quantifying how this advice/assistance provided by CTR contributes to the overall availability of the facility to perform its agent destruction mission is the challenge. Two quantifiable factors that can be identified and tracked from a strategic perspective are the amount of funding invested in providing technical assistance and the amount of agent and munitions destroyed by the facility.

Although it can be quantified, tracking funds expended on technical assistance provides no clear measure of whether the technical assistance actually is contributing to the facility's ability to destroy nerve agent. Costs are incurred whether or not any deficiency is successfully identified, isolated, and corrected. However, the number of munitions processed and the tonnage of nerve agent destroyed does provide a valuable and useful link to determining how successful the provided technical assistance has been toward keeping the facility functioning at capacity. Progress in the destruction of agents and processing of munitions is clearly indicative of continued facility functionality and operation. Therefore, we determined that using the amount of agent destroyed and tracking the number of munitions processed would be the best metrics to measure the success of the technical support presently being provided.

Description of New Program-Level Metrics

As nerve agent-filled munitions are brought to the facility for processing, the weight of nerve agent extracted and neutralized is determined and the number of individual munitions bodies processed is captured by the plant's automated process control system. The Russian Federation's operations staff consolidates and reports the two quantities, and the U.S. on-site technical staff obtains the data from the Russian Federation and maintains a cumulative total of both data points. The on-site staff then provides the information weekly to the U.S. Project Management Office.

Table 1 below provides an example of the metrics provided in the weekly report:

Table 1. CWE Metrics

Report Date	Total Rounds	Total Metric Tons
August 3, 2010	532,806	1461.8

However, since the number of munitions (total rounds) correlates closely to the amount of nerve agent destroyed, for the purposes of reporting externally in the "CTR scorecard" only the metric tons of agent destroyed will be reported. (Appendix 1),

Project-Level Metrics

Additional metrics will be maintained at a project level to assess how effectively the destruction facility is being used in relation to performance of maintenance and repairs that help ensure the safety of the destruction process.

Methodology for Development of New Project-Level Metrics

Reliability, Availability, and Maintainability (RAM) Analysis

When fully constructed, the Shchuch'ye CWDF will consist of two main munitions destruction buildings, with two destruction process lines (DPLs) operating in each building. A RAM analysis to establish the baseline threshold and objective capacities of the facility in this configuration was completed in August 2005. That analysis determined that the design capacity ***threshold*** for the completed facility is 1,000 metric tons of agent destruction per year, with a design capacity ***objective*** of 1,700 metric tons of agent destruction per year (capacity for two main destruction buildings with two DPLs operating in each).

Description of new project-level metrics:

Scheduled Facility Downtime

Scheduled Facility Downtime (SFD) tracks any days when it is planned that the facility will be idle and not processing munitions. This can be due to scheduled maintenance, campaign changeovers, or any administrative downtime where the idling of the process was planned in advance. In the RAM analysis, the SFD was estimated to average 25 percent. For a facility of this complexity, with its associated preventive maintenance requirements both for the destruction process and all of the support systems, such an estimate is reasonable.

Minimizing down days, whether scheduled or unscheduled, is extremely important. Using scheduled downtime efficiently and effectively by scheduling periodically required maintenance actions for those periods can maintain overall production capabilities, as well as the safety of the facility. U.S. technical support efforts help the Russian Federation to plan and optimize scheduled maintenance outages and helps ensure that the scheduled outage does not extend beyond the originally scheduled duration by having the right parts, equipment and vendor support pre-staged to take full advantage of the outage.

Since the start of munitions destruction, with only one of the two destruction buildings in operation, the cumulative SFD has been 19.25%. A rolling 12-month average helps identify potential trends that might get masked over the long term. The Shchuch'ye CWDF's current rolling 12-month SFD average is 26.85%. This higher percentage is attributable to the recent scheduled changeover from one type of munition to another.

In terms of the objectives and thresholds for agent tons per year, an objective SFD average of 20 percent per year would support meeting the 1,700 metric tons of agent per year average with two main production facilities operational. Similarly, an SFD average of 35 percent would result in meeting the threshold average of 1,000 metric tons per year.

Unscheduled Facility Downtime

A second key area for assessing overall throughput is Unscheduled Facility Downtime (UFD). Efficient and effective technical support helps minimize the number of UFD days. This support encompasses proactive failure root-cause analysis, analysis of spare parts consumption, provisioning of the right amount and types of spare parts, and the selective replacement of items before failure during scheduled maintenance outages.

Objective and threshold limits for this metric are being developed.

The Shchuch'ye CWDF's current overall cumulative UFD percentage is 6.68%, with a 12-month rolling average of 4.66%. The UFD for the Shchuch'ye CWDF is significantly lower than would be expected for a complex CWDF, based upon experience with similar facilities in the U.S. program, especially during the initial year of operation.

Facility Achieved Availability

The Achieved Availability metric encompasses the overall throughput of the facility achieved versus the actual design capacity of the facility. During the current phase of operations, it is based upon the one completed main destruction building (overall capacity based upon two DPLs). The achieved availability with one main destruction building is a good indicator of what could be expected with two main destruction buildings in terms of meeting the overall objective of 1,700 metric tons per year for a facility operating with two main destruction facilities.

Based upon the RAM analysis, the estimated Achieved Availability was 35.4 percent, which would result in an overall annual throughput rate of 1,382.5 metric tons for the facility with two main destruction buildings in operation with two operational DPLs in each. In order to achieve the objective of 1,700 metric tons per year, an overall Achieved Availability of approximately 44 percent would be required. Similarly, it is estimated that the 1,000 metric tons per year threshold would be met with an Achieved Availability of 26 percent. To date, Building 101A has an Achieved Availability of 54 percent, much higher than originally anticipated.

The Achieved Availability calculation takes into account Scheduled and Unscheduled Facility Downtimes. Based on operations to date, these two components account for almost 26 percent of the nearly 45 percent of lost production time for Building 101A associated with an Achieved Availability of 54 percent. The remaining 19 percent of downtime is primarily due to unplanned and planned maintenance of one or the other of the two DPLs while one remains in operation.

Summary

We are using three project-level metrics to track and evaluate the trend of the effectiveness of the facility and the provided technical support. Based upon an analysis of available RAM reports, the following are the project-level metrics, with associated objective and threshold values, as of June 30, 2010:

Scheduled Facility Downtime (SFD):
Objective: 20 percent (RAM analysis rate of 1,700 metric tons/year)
Threshold: 35 percent (RAM analysis rate of 1,000 metric tons/year)

Cumulative SFD: 19.25 percent
Last 12 Months SFD: 26.85 percent

Unscheduled Facility Downtime (UFD):
Cumulative UFD: 6.7 percent
Last 12 Months UFD: 4.7 percent

Facility Achieved Availability (FAA):
Objective: 44 percent
Threshold: 26 percent

Cumulative FAA: 53.97 percent
Last 12 Months FAA: 52.51 percent

Nuclear Weapons Safety and Security

Program Description

This program supports proliferation prevention by enhancing the security systems of nuclear weapons storage sites using DoD nuclear security standards as a basis for design. It also trains cadres of security systems operators, administrators, and maintenance and repair technicians and provides capabilities, such as the personnel reliability program and guard force training systems. In response to a request from Russia and former President Bush's commitment at Bratislava, Slovakia in February 2005, DoD and the Department of Energy (DOE) enhanced security systems at requested locations that permanently or temporarily store strategic and non-strategic (tactical) nuclear weapons. DoD upgraded a total of 24 sites and, additionally, supports proliferation prevention by enhancing the security and safety of nuclear weapons during shipment through the Nuclear Weapons Transportation Security (NWTS) Program.

Methodology for Development of New Metrics

Nuclear Weapons Safety and Security (NWS&S) projects have been focused on efforts exclusively in Russia, and the metric used to measure program success reported the number of nuclear weapons storage sites upgraded. This remains an appropriate measure of the impact and effectiveness of program efforts in Russia and, therefore, this information will continue to be reported on the CTR Scorecard (Appendix 1) in its present format.

Recently, however, with the potential global expansion of the Nuclear Weapons Storage Security (NWSS) program to other partner countries, DTRA has been working to develop enhanced metrics that better reflect how our efforts are contributing to overall threat reduction. These new metrics are aimed at measuring the capability of Russia (and in the future, other countries) to sustain the ability to store and transport nuclear materials safely and securely. The development of these metrics presents a substantial challenge, as we need to develop measures that accurately and correctly reflect an intangible capability. DoD and DOE have been working together on this issue and have jointly developed eight principles to evaluate sustainment capabilities. Although these principles are presently used as internal measures of effectiveness and are still being refined, the NWSS program has already begun to apply these principles to the sustainment of the site security projects in Russia.

Description of New Metrics

DoD and DOE jointly developed eight sustainment and logistics principles for the Russia NWSS projects, and proposed metrics would follow these standards. These metrics also meet other international standards, including those used by the International Atomic Energy Agency (IAEA) for logistics and maintenance management of nuclear facilities. The metrics include:

- Performance/Capability Assurance,
- Configuration Management,
- Procedures and Processes,

- Training,
- Organization and Personnel,
- Life Cycle Management,
- Maintenance, and
- Logistics.

These principles/goals are further defined in Table 3 below with the associated objectives and the metrics being tracked. Further development of the methodology is planned jointly with DOE to clarify roles and responsibilities more completely. These metrics would be provided as an appendix to the CTR Scorecard

Note: NWTS program metrics will remain the same, as these continue to track the threat reduction impact of this project. The number of railcar shipments will continue to be included in the scorecard and will measure the number of nuclear weapon transport shipments supported annually. An additional metric will be added to the CTR Scorecard to reflect the number of new nuclear weapons guard and cargo railcars with enhanced security systems and off-train communications provided.

TABLE 3.1 NWSS Goals, Objectives, and Metrics

Goals	Objectives	Metrics
1.0 Performance/Capability Assurance	1.1 Partner Country institutes an assessment program, including vulnerability assessments and system readiness tests, to ensure site security effectiveness is sustained.	• Metrics to fulfill this goal are still being determined based on current sustainment activities. *This goal is challenging, as it relates to a host country's ability to self-assess its performance in the below areas. Depending on the level of cooperation, DoD may not be able to track this metric independently. The metric is included, however, as it has been shared with and agreed to by the host country to assist it in taking over sustainment activities once U.S. assistance stops.*
2.0 Configuration Management	2.1 Partner Country institutes a program to maintain configuration control for all site security elements.	• Functional technical library exists. • Equipment lists and design specifications by site are maintained.
3.0 Procedures and Process	3.1 Partner Country institutes systems, processes and procedures to ensure site security effectiveness is sustained.	• Metrics to fulfill this goal are still being determined based on current sustainment activities. *This goal is challenging, as it relates to a host country's internal procedures and processes. Depending on the level of cooperation, DoD may not be able to track this metric independently. The metric is included, however, as it has been shared with and agreed to by the host country to assist it in taking over sustainment activities once U.S. assistance stops.*
4.0 Training	4.1 Partner Country institutes a training program and delivers the majority of training	• Number of personnel trained o Operators o Administrators o Maintainers

Goals	Objectives	Metrics
	necessary through an organic capability to operate and maintain security elements.	• Establishment or enhancement of regional technical and training centers • Functional technical library exists.
5.0 Organization and Personnel	5.1 Partner Country staffs sustainment centers and sites with personnel qualified to sustain operations and maintenance of security elements.	• Quantity of trained Physical Protection System Operators, Administrators and Maintainers/ Repair Technicians
6.0 Life Cycle Management	6.1 Partner Country regularly assesses system reliability, availability and maintainability of security elements, using usage and failure data to plan and implement improvements and upgrades to sustain security effectiveness.	• Failure data is tracked in order to determine trends in equipment life cycle. • Partner country has spares and consumables available and a preventive management system in place.
7.0 Maintenance	7.1 Partner Country ensures that an organic capability exists to perform the majority of maintenance and repair tasks to sustain security elements.	• Failure data reports generated and provided to DoD indicating occurrence of maintenance activity, failures, and successful repair/replacement of failed parts/systems. • Functional technical library exists
8.0 Logistics	8.1 Partner Country performs all necessary logistics tasks to sustain security elements.	• Necessary equipment has been procured and delivered, is available and is operational. Overall security systems are functioning effectively. • Reports received on the amounts of support equipment /spares provided, annual usage rate, and fill rate.

Weapons of Mass Destruction Proliferation Prevention Program

Program Description

The continued threat from possible trafficking of WMD and related materials demands an expansion of and a new approach to proliferation prevention efforts. The WMD Proliferation Prevention Program (WMD-PPP) is focused on enhancing partner country border security capabilities to prevent proliferation and will assist those countries in the development of effective and sustainable capabilities to prevent the proliferation of WMD, related materials, and technologies threatening U.S. national security and global stability.

WMD-PPP has evolved from an "initiative" into an established program and will apply the "lessons learned" as the program expands globally. This expansion requires a new vision, expanded mission, and associated strategic goals with a global focus on stopping the movement of WMD. With its unique expertise in enhancing deterrence, detection, and interdiction capabilities on land or at sea, the WMD-PPP is capable of expanding its reach beyond individual countries to include regional and global partnerships. This program is able to provide a system of materiel and non-materiel solutions tailored to the needs of the partner state to counter the threat of WMD smuggling. Furthermore, by ensuring that the assistance provided can be effectively sustained by our partners, we are able to be confident that our investment in proliferation prevention solutions will be long lasting.

Methodology for Development of WMD-PPP Metrics

To determine measures of success, metrics have been developed to address three strategic program goals. Specific objectives for each goal are listed under the program goal. They are as follows:

1. Strengthen the capability of partner countries to secure their borders against illicit movement of WMD, related materials, and technologies.
    a. Enhance partner country capability to perform effective risk management.
    b. Enhance partner country capability to perform border security command, control, communications, and computers (C4).
    c. Enhance partner country capability to perform border security surveillance.
    d. Enhance partner country capability to perform WMD detection.
    e. Enhance partner country capability to interdict border violations

2. Assist partner countries to achieve long-term sustainment of border security capabilities.
    a. Assist partner country with the development of a sustainment budget for all systems delivered under this program.
    b. Enhance partner country capability to support and maintain delivered equipment and/or systems.
    c. Enhance partner country capability to sustain delivered training.

3. Promote cooperative relationships to further U.S. national security and global stability.

a. Enhance partner country capability to capture and disseminate information concerning WMD incidents to applicable USG agencies.
b. Increase the awareness of partner countries as to their critical role in WMD proliferation prevention.
c. Develop bilateral, regional, and multilateral cooperation.

Program Metrics

The ability to measure simply and objectively the impact that WMD-PPP assistance has had on threat reduction is challenging due to the nature of the program: we are providing a capability to our partners that gives them an ability to deter proliferation. Thus, as we are unable to show quantitatively that proliferation has, in fact, been prevented, we must measure the efficiency and effectiveness of the capacity provided. It is possible to quantify the amount of work the WMD-PPP is accomplishing in efforts to provide this capability indirectly by measuring and reporting the following metrics:

- Miles of green and blue border with additional security enhancements
- Number of ports of entry where capabilities have been enhanced

On an annual basis, the WMD-PPP will provide updated totals for the above two metrics, which henceforth will appear on the CTR Scorecard.

Objective quantification of the impact of these enhancements is more difficult to establish. Although data can be gathered on the number of detected and interdicted attempts to cross the border, it is impossible to know what number of attempts have not been detected. Furthermore, enhancing deterrence can reduce the number of detections and interdictions that would otherwise have been made. In order to overcome this challenge, one way to measure the effectiveness of enhancements objectively is through testing; that is, standardized exercises that can be conducted before and after enhancements are made to measure the impact of those enhancements.

Project-Level Metrics

The table below provides a description of new project-level metrics. These lower-level metrics provide greater detail as to the impact of each individual project and can be provided as an addendum to the simpler Scorecard metrics.

TABLE 4. WMD-PPP Goals, Objectives, and Metrics

Goals	Objectives	Metrics
1.0 Strengthen the capability of partner countries to secure their borders against illicit movement of WMD, related materials, and technology.	1.1 Enhance partner country capability to perform effective risk management (RM)	• Percentage of alerts from RM system that result in a positive outcome • Increase in efficiency of RM system • Increase in effectiveness of RM system (percentage of alerts that result in a positive outcome) • Number of alerts provided to operational units • Number of independent sources integrated into RM system

Goals	Objectives	Metrics
	1.2 Enhance partner country capability to perform border security command, control, communications and computers (C4)	• Number of border security facilities provided with enhanced C4 • Number of border stations and ports of entry provided with enhanced communications • Provided partner country capability to "manage" or "control" X number of maritime ports or X miles of border • Number of command centers connected to lower level units • Number of sensor and information types integrated into command & control (C2) system
	1.3 Enhance partner country capability to perform border security surveillance	• Provided partner country capability to surveil/monitor X number of miles/km of maritime or land border • Nautical miles of shipping lanes surveilled
	1.4 Enhance partner country capability to perform WMD detection	• Number of border facilities (land and maritime) provided increased capability to detect WMD • Number of personnel trained and equipped to perform WMD detection
	1.5 Enhance partner country capability to perform border security interdiction	• Number of personnel trained and equipped to perform border security interdiction • Number of border units (border stations and vessels) provided improved capability to perform interdiction • Enhanced X number of interdiction platforms • Number of boarding teams trained and equipped to perform maritime interdiction
2.0 Assist partner countries to achieve long-term sustainment of border security capabilities	2.1 Assist partner country with the development of a sustainment budget for all systems delivered under this program	• Percentage of sustainment cost budgeted by partner country for all systems delivered under this program • Percentage of sustainment cost funded by partner country
	2.2 Enhance partner country capability to support and maintain delivered equipment and/or systems	• Number of maintenance personnel trained • Number of support equipment sets provided • Operational availability of delivered equipment and/or systems • Does partner country have spares and consumables available and existing preventive maintenance system? • Functional technical library
	2.3 Enhance partner country capability to sustain delivered training	• Initial and refresher training system exists • Current operators/maintainers trained • Establish training centers • Number of mobile training teams • Established training records • Training materials (programs of instruction, computer based training , manuals, etc) • Training and certifications maintained • Incorporation of training in existing doctrine and academies
3.0 Promote cooperative relationships to further	3.1 Enhance a partner country capability to	• Number of countries in which we have established reporting mechanisms for WMD

Goals	Objectives	Metrics
U.S. national security and global stability	capture and disseminate information including that concerning WMD incidents	incidents (means, procedure, agreement to share)
	3.2 Increase the awareness of partner countries as to their critical role in WMD non-proliferation (partner country buy-in)	• Number of personnel exchanges (training, professional development) • Increase in partner country's commitment to adhere to UNSCR 1540 and other non-proliferation regimes
	3.3 Develop interagency, bilateral, regional, and multilateral cooperation	• Mean time to negotiate agreements • Number of regional relationships facilitated • Number of regional/multilateral exercises

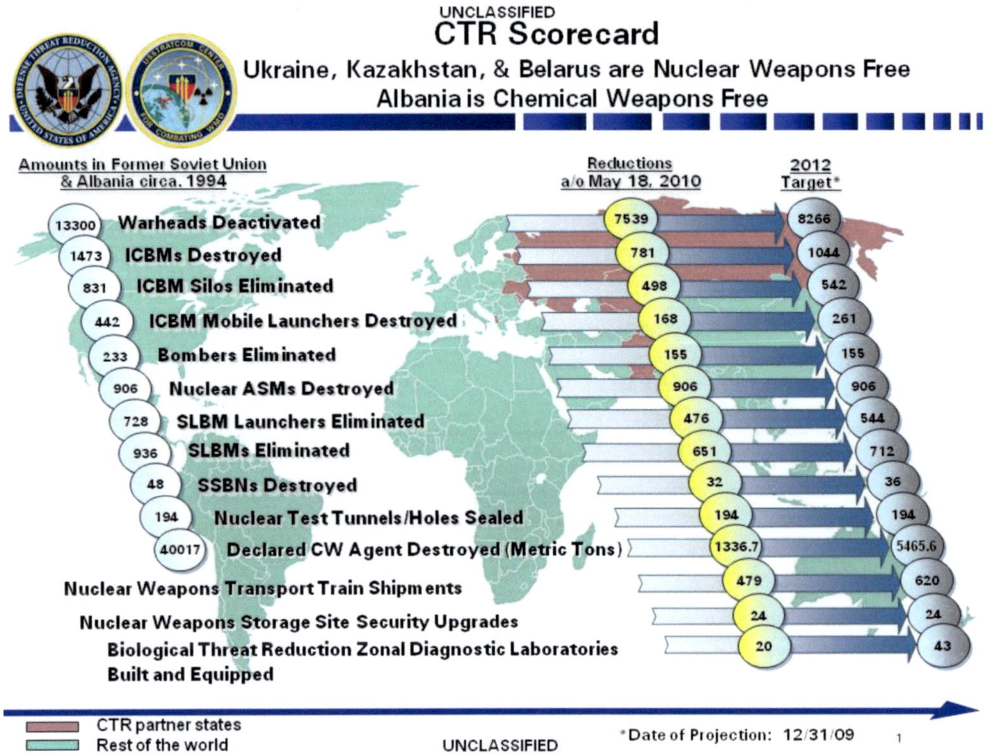
UNCLASSIFIED
CTR Scorecard
Ukraine, Kazakhstan, & Belarus are Nuclear Weapons Free
Albania is Chemical Weapons Free
Amounts in Former Soviet Union & Albania circa. 1994
Reductions a/o May 18, 2010
2012 Target*
13300 Warheads Deactivated 7539 8266
1473 ICBMs Destroyed 781 1044
831 ICBM Silos Eliminated 498 542
442 ICBM Mobile Launchers Destroyed 168 261
233 Bombers Eliminated 155 155
906 Nuclear ASMs Destroyed 906 906
728 SLBM Launchers Eliminated 476 544
936 SLBMs Eliminated 651 712
48 SSBNs Destroyed 32 36
194 Nuclear Test Tunnels/Holes Sealed 194 194
40017 Declared CW Agent Destroyed (Metric Tons) 1336.7 5465.6
Nuclear Weapons Transport Train Shipments 479 620
Nuclear Weapons Storage Site Security Upgrades 24 24
Biological Threat Reduction Zonal Diagnostic Laboratories Built and Equipped 20 43
CTR partner states
Rest of the world
UNCLASSIFIED
*Date of Projection: 12/31/09
1

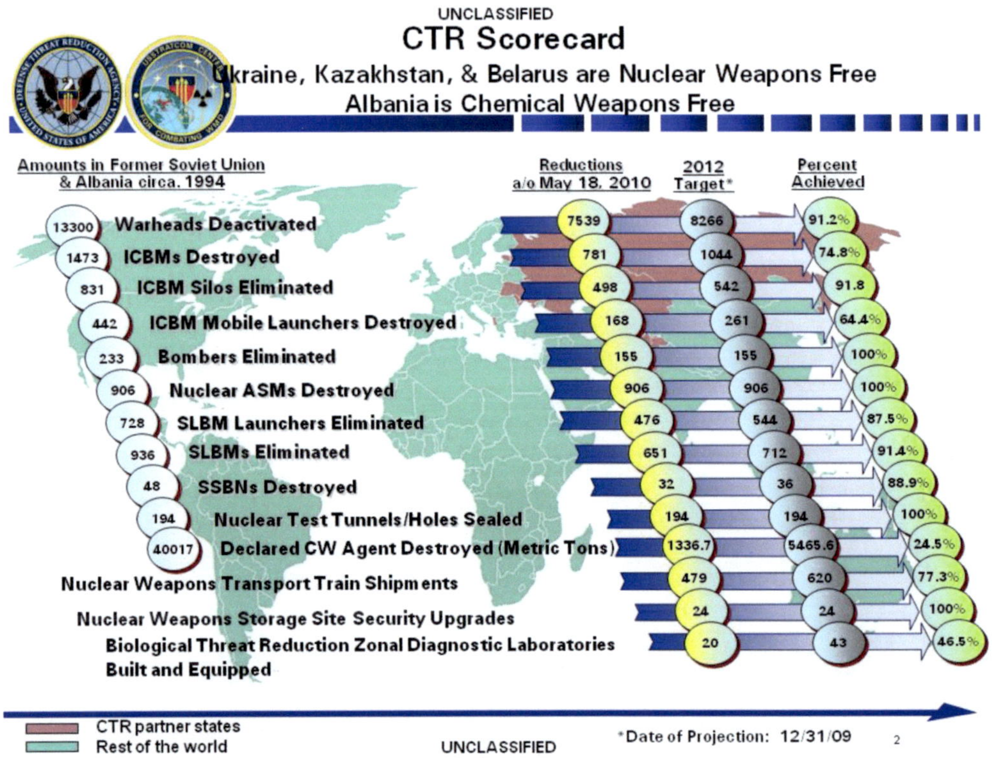
UNCLASSIFIED
CTR Scorecard
Ukraine, Kazakhstan, & Belarus are Nuclear Weapons Free
Albania is Chemical Weapons Free
Amounts in Former Soviet Union & Albania circa. 1994
Reductions a/o May 18, 2010
2012 Target*
Percent Achieved
13300 Warheads Deactivated 7539 8266 91.2%
1473 ICBMs Destroyed 781 1044 74.8%
831 ICBM Silos Eliminated 498 542 91.8
442 ICBM Mobile Launchers Destroyed 168 261 64.4%
233 Bombers Eliminated 155 155 100%
906 Nuclear ASMs Destroyed 906 906 100%
728 SLBM Launchers Eliminated 476 544 87.5%
936 SLBMs Eliminated 651 712 91.4%
48 SSBNs Destroyed 32 36 88.9%
194 Nuclear Test Tunnels/Holes Sealed 194 194 100%
40017 Declared CW Agent Destroyed (Metric Tons) 1336.7 5465.6 24.5%
Nuclear Weapons Transport Train Shipments 479 620 77.3%
Nuclear Weapons Storage Site Security Upgrades 24 24 100%
Biological Threat Reduction Zonal Diagnostic Laboratories Built and Equipped 20 43 46.5%
CTR partner states
Rest of the world
UNCLASSIFIED
*Date of Projection: 12/31/09
2

Appendix 2: CBEP Objective, Capability and MOE Traceability

CBEP Objective	CBEP Capability	Attribute	Measure of Effectiveness (MOE)	Indicator	Evaluation Criteria
CBEP Objective 1: Secure and consolidate collections of especially dangerous pathogens (EDP) and their associated research at a minimum number of secure health and agricultural laboratories or related facilities.	**Capability 1:** Secure and consolidate collections of EDPs and their associated research.	*Consolidation*	**MOE 1:** Partner country EDP collections[i] and associated research are consolidated into a minimum number[ii] of locations.[iii]	**1.1:** Number of EDP collections.	**Min**: No increase in the number of EDP collections over baseline[iv]. **Obj**: No more than 1 EDP collection.
		Security	**MOE 2:** Partner country EDPs and associated research are secured in a manner consistent with standards.	**2.1:** Biosecurity compliance.	**Min**: Consistent with the European Committee for Standardization (CEN) Workshop Agreement (**CWA**) Biorisk Management requirements. **Obj**: Consistent with the requirements of DoDI 5210.89 outlined in CTB-09-405 Memorandum "Biological Threat Reduction Program (BTRP) Biosafety and Security Standards (BS&S) ," dated November 20, 2009 (***CTB-09-405 Memorandum***).
CBEP Objective 2: Enhance partner country/region's capability to prevent the sale, theft, diversion, or accidental release of biological weapons (BW)-related materials, technology, and expertise by improving biological safety and security (BS&S) standards and procedures.	**Capability 1:** Secure and consolidate collections of EDP and their associated research.	*Consolidation*	**MOE 1:** Same as Objective 1, Capability 1 MOE 1	**1.1:** Same as Objective 1, Capability 1 MOE 1.	**Min**: Same as Objective 1, Capability 1 MOE 1. **Obj**: Same as Objective 1, Capability 1 MOE 1.
		Security	**MOE 2:** Same as Objective 1 Capability 1 MOE 2	**2.1:** Same as Objective 1, Capability 1 MOE 2	**Min**: Same as Objective 1, Capability 1 MOE 2. **Obj**: Same as Objective 1, Capability 1 MOE 2.
	Capability 2: Enhance partner country BS&S standards and practices.	*Enforceability*	**MOE 3:** Partner country has BS&S laws and regulations governing work with EDPs.	**3.1:** Established legal framework[v].	**Min**: Partner country laws, guidelines, and regulations demonstrate commitment and progress towards compliance with international laws, regulations, and treaties (BWC, WHO, OIE, UN, etc.). **Obj**: Partner country laws, guidelines, and regulations are compliant with DODI 5210.89 to the maximum extent possible and CDC/NIH "Biosafety in Microbiological and Biomedical Laboratories, 5th edition, 2007" (***BMBL 5***) and subsequent versions.

CBEP Objective	CBEP Capability	Attribute	Measure of Effectiveness (MOE)	Indicator	Evaluation Criteria
				3.2: Level of Regulation.	<u>**Min**</u>: Compliance at the national level. <u>**Obj**</u>: Compliance at the local, regional, and national levels.
		Safety	**MOE 4:** Partner country guidelines for work with EDPs meet U.S. or international guidelines for biosafety.	**4.1:** Biosafety guidelines.	<u>**Min**</u>: Consistent with the minimum requirements of WHO IHR. <u>**Obj**</u>: Consistent with the requirements of ***BMBL 5*** and subsequent editions for safety in accordance with ***CTB-09-405 Memorandum***.
				4.2: Facility specific biosafety plans exist.	<u>**Min**</u>/<u>**Obj**</u>: Yes.
		Security	**MOE 5:** Partner country standards for work with EDPs are consistent with standards for biosecurity.	**5.1:** Biosecurity standards.	<u>**Min**</u>: Consistent with the CWA Biorisk Management requirements. <u>**Obj**</u>: Consistent with the requirements of DoDI 5210.89 outlined in ***CTB-09-405 Memorandum***.
				5.2: Facility specific biosecurity plans exist.	<u>**Min**</u>/<u>**Obj**</u>: Yes.
		Transparency	**MOE 6:** Partner country BS&S standards and practices are transparent.	**6.1:** BS&S standards are available.	<u>**Min**</u>: Available to independent third party. <u>**Obj**</u>: Available to U.S. Government.
				6.2: Biosecurity Event Notification.	<u>**Min**</u>/<u>**Obj**</u>: Immediate notification of the theft or loss of an EDP to the appropriate Ministry and law enforcement/security organizations, as well as in accordance with international obligations and treaties.
				6.3: Biosafety Event Notification.	<u>**Min**</u>/<u>**Obj**</u>: Immediate notification of the release of an EDP causing occupational exposure or release of an EDP outside of the primary barriers of the biocontainment[vi] area to the appropriate Ministry and law enforcement/security organizations, as well as in accordance with international obligations and treaties.

CBEP Objective	CBEP Capability	Attribute	Measure of Effectiveness (MOE)	Indicator	Evaluation Criteria
CBEP Objective 3: Enhance partner country/region's capability to detect, diagnose, and report endemic and epidemic [vii], man-made or natural EDPs, bio-terror attacks, and potential pandemics.	**Capability 1:** Enhance partner country disease detection, diagnosis [viii], and reporting capabilities.	*Safety*	**MOE 1:** Partner country disease detection and diagnosis capability meets U.S. and/or international guidelines for biosafety.	**1.1:** Biosafety guidelines	**Min**: Consistent with the minimum requirements of WHO IHR. Obj: Consistent with the requirements of ***BMBL 5*** and subsequent editions for safety in accordance with ***CTB-09-405 Memorandum***.
		Security	**MOE 2:** Partner country disease detection[ix] and diagnostic infrastructure complies with standards for biosecurity.	**2.1:** Biosecurity standards	**Min**: Consistent with the CWA Biorisk Management requirements. Obj: Consistent with the requirements of DoDI 5210.89 outlined in ***CTB-09-405 Memorandum.***
		Comprehensive	**MOE 3:** Partner country has preparedness and response plans.	**3.1:** National Pandemic Influenza Preparedness and Response plan[x].	**Min**: National Pandemic Influenza Preparedness and Response Plan has been developed. **Obj**: Plan conforms to WHO guidance document for Pandemic Influenza Preparedness and Response.
				3.2: Bioterrorism preparedness and response plan.	**Min**: Bioterrorism preparedness and response are covered in a developed Public Health Emergency/Bioterrorism Response Plan. **Obj**: Plan conforms to HHS/CDC template[xi].
				3.3: Multi-Hazard [xii] National Public Health Emergency Preparedness and Response Plan.	**Min**: Multi-hazard National Public Health Emergency Preparedness and Response Plan for Biological Hazards has been developed. **Obj**: Plan conforms to core biological capacity requirements IAW WHO/IHR[xiii] Monitoring Framework.
				3.4: National Emergency Preparedness and Response Plan for animal diseases.	**Min**: National Emergency Preparedness and Response Plan for animal diseases has been developed. **Obj**: Plan conforms to FAO/OIE manual[xiv].

CBEP Objective	CBEP Capability	Attribute	Measure of Effectiveness (MOE)	Indicator	Evaluation Criteria
		Comprehensive	**MOE 4:** Partner country understands its EDP disease and pathogen baselines.	**4.1:** National disease surveillance plan.	**Min**: National disease surveillance plan exists. **Obj**: National active and passive surveillance[xv] is established and executed IAW WHO Recommended Surveillance Standards, Second Edition.
		Comprehensive	**MOE 5:** Partner country disease surveillance system is capable of detecting and reporting suspect EDP cases to those responsible for human and animal health.	**5.1:** Disease surveillance system is capable of detecting EDP cases.	**Min**: Disease surveillance system is capable of detecting endemic EDP suspect cases. **Obj**: Disease surveillance system is capable of detecting all suspect EDP cases.
				5.2: Epidemiological data from EDP case investigations are shared with those responsible for human and animal health.	**Min/Obj**: Epidemiological EDP data from case investigations are shared with those responsible for human and animal health within partner country's own government structure.
				5.3: Laboratory test results are provided to those responsible for human and animal health.	**Min/Obj**: Laboratory test reports from EDP cases are shared with those responsible for human and animal health within partner country's own government structure.
				5.4: All reported human epidemiological events constituting a Public Health Emergency of International Concern (PHEIC) are reported to the WHO[xvi].	**Min/Obj**: All events constituting a PHEIC are reported to WHO IAW IHR regulations.
				5.5: All animal reportable diseases are reported to OIE.[xvii]	**Min/Obj**: All events constituting OIE animal reportable diseases are reported to OIE IAW OIE guidelines.
				5.6: Case data is shared via appropriate reporting systems.	**Min/Obj**: Reporting systems meet WHO[xviii] and OIE[xix] standards for early detection and timely reporting.

CBEP Objective	CBEP Capability	Attribute	Measure of Effectiveness (MOE)	Indicator	Evaluation Criteria
		Timeliness	**MOE 6:** Partner country disease surveillance system is capable of early detection and timely reporting of EDP cases to those responsible for human and animal health.	**6.1:** Epidemiological data from EDP case investigations are promptly reported to those responsible for human and animal health.	**Min/Obj**: EDP case investigation reports are shared within 24 hours[xx] of investigation.
				6.2: Laboratory test results are promptly provided to those responsible for human and animal health.	**Min/Obj**: Laboratory test results are shared within 24-hours.
				6.3: All human epidemiological events constituting a Public Health Emergency of International Concern (PHEIC) are promptly reported to the WHO.[xxi]	**Min/Obj**: All PHEIC reports are shared IAW WHO IHR timelines.
				6.4: All animal reportable diseases are promptly reported to the OIE.[xxii]	**Min/Obj**: All reportable animal disease reports are shared with OIE IAW OIE timelines.
		Comprehensive	**MOE 7:** Partner country is capable of investigating and diagnosing EDP cases.	**7.1:** Suspect EDP cases are investigated and documented by those responsible for human or animal health.	**Min**: Standardized case investigation form is completed and complies with WHO/OIE standard. **Obj**: Standardized case investigation form is completed in an electronic system and complies with WHO/OIE standard.[xxiii]
				7.2: Appropriate samples are collected and promptly trans-ported under optimum conditions[xxiv] for	**Min/Obj**: Samples are collected and transported IAW CBEP procedures.

CBEP Objective	CBEP Capability	Attribute	Measure of Effectiveness (MOE)	Indicator	Evaluation Criteria
				laboratory confirmation of the diagnosis.	
				7.3: Partner country is able to diagnose endemic EDPs.	**Min**: Diagnosis is made IAW CBEP procedures and demonstrated through proficiency testing program. **Obj**: Diagnosis is made IAW CBEP procedures and demonstrated in country during real time case investigation and diagnosis.
				7.4: Partner country is able to utilize international reference laboratories when there is no country diagnostic capability.	**Min**: Capable of shipping samples IAW IATA (International Airline Transport Association) regulations and requirement agreements of the international reference lab. **Obj**: Established agreement exists within international reference labs to which samples will be routinely shipped.
		Timeliness	**MOE 8:** Suspect EDP cases are promptly investigated and diagnosed.	**8.1:** Suspect cases are promptly investigated and assessed by those responsible for human or animal health.	**Min/Obj**: Within 48-hours of suspect case being reported to those responsible for human and animal health IAW WHO IHR/OIE guidelines.[xxv]
				8.2: Appropriate samples are collected and promptly transported under optimum conditions for laboratory confirmation of the diagnosis.	**Min**: Initiate sample transport within 24-hours of case investigation. **Obj**: Initiate sample transport immediately following case investigation.
				8.3: Partner country promptly initiates diagnostic testing of endemic EDPs.	**Min**: Initiate appropriate diagnostic testing protocols/SOPs within 24-hours of receipt of samples. **Obj**: Initiate appropriate diagnostic testing protocols/SOPs immediately upon receipt of samples.
				8.4: Partner country is able to rapidly	**Min/Obj**: Ship samples to international reference laboratory within 24-hours of

CBEP Objective	CBEP Capability	Attribute	Measure of Effectiveness (MOE)	Indicator	Evaluation Criteria
				utilize international reference laboratories when there is no in-country diagnostic capability.	acquiring samples.
CBEP Objective 4: Ensure the developed capabilities are designed to be sustainable within each partner country/region's current operating budget.	**Capability 1:** Secure and consolidate collections of EDP and their associated research.	*Sustainability*	**MOE 1:** Partner country EDP collection methods and associated research are sustainable.	**1.1:** Sustainability plan to physically maintain collection.	**Min/Obj**: Plan identifies human, resource and operational needs to maintain collection of viruses and bacteria over time to support pathogen baseline and facilitate U.S. efforts to mitigate biological risks.
	Capability 2: Enhance partner country BS&S standards and practices.	*Sustainability*	**MOE 2:** Partner country BS&S practices and regulations concerning EDPs are absorbable with respect to cost.	**2.1:** Sustainment cost of application and enforcement of BS&S guidelines and regulations.	**Min/Obj**: Cost is within minimally budgeted resources of the partner country and in partnership with others.
	Capability 3: Enhance partner country disease detection, diagnosis, and reporting capabilities	*Sustainability*	**MOE 3:** Partner country disease detection, diagnosis, and reporting capabilities are absorbable with respect to cost.[xxvi]	**3.1:** Sustainment Cost.	**Min/Obj**: Cost is within minimally budgeted resources of the partner country and in partnership with others.
	Capability 4: Enhance partner country ability to train/retain personnel trained to desired biosafety/ biosecurity standards.	*Sustainability*	**MOE 4:** Partner country demonstrates an ability to retain and/or train qualified personnel to desired standards.	**4.1** Trainee test results.	**Min/Obj:** 70% of trained biosafety/biosecurity officers can demonstrate continued proficiency (by means of either written or hands-on testing) at 6, 12, and 18 months following completion of training.

CBEP Objective	CBEP Capability	Attribute	Measure of Effectiveness (MOE)	Indicator	Evaluation Criteria
CBEP Supplemental Objective 5: Facilitate the engagement of partner country's/regional scientific and technical personnel in research areas of interest to both the partner country/region and the United States.	**Capability 1:** Engage partner country life science personnel in areas of interest to partner country/ region and the United States.	*Legitimacy*	**MOE 1:** Research objectives of partner country scientists and institutes are aligned with national and international EDP priorities.	**1.1:** Credible, appropriate and executable EDP research projects.	**Min**: CBEP sponsored research projects are technically reviewed and approved by CBEP SMEs. **Obj**: Country projects on EDPs are competitively peer reviewed and funded by international review board.
		Transparency	**MOE 2:** Partner country scientists and institutes openly share research information and copies of EDP strains.	**2.1:** U.S. access to research data funded by CBEP.	**Min/Obj**: U.S. has access to all data generated by joint research projects.
				2.2: Copies of EDP strains transferred to U.S.	**Min**: Copies of all requested EDP strains from joint research projects are transferred to the U.S. **Obj**: Copies of all requested EDP strains are transferred to the U.S.
				2.3: Contribution to international scientific community.	**Min**: Partner country scientists present BTRP research results at relevant international conferences, and contribute to research articles/presentations published/presented within the international community. **Obj**: Partner country scientists are invited to present their research results at relevant international conferences, and have lead role in publishing research articles and/or conducting presentations within the international community.
		Safety	**MOE 3:** Work of partner country scientists and institutes with EDPs meets U.S. or international guidelines for biosafety.	**3.1:** Biosafety guidelines.	**Min**: Consistent with the minimum requirements of WHO IHR. **Obj**: Consistent with the requirements of ***BMBL 5*** and subsequent editions for safety in accordance with ***CTB-09-405 Memorandum***.
		Security	**MOE 4:** Partner country scientists and institutes work with EDPs is consistent with standards for biosecurity.	**4.1:** Biosecurity standards.	**Min**: Consistent with the CWA Biorisk Management requirements. **Obj**: Consistent with the requirements of DoDI 5210.89 outlined in ***CTB-09-405 Memorandum***.
CBEP Supplemental Objective 6:	**Capability 1:** Eliminate bioweapons	*Comprehensive*	**MOE 1:** Disclosed BW-related infrastructure, materials, and technologies are eliminated.	**1.1:** Extent of elimination	**Min**: Scale and scope consistent with peaceful purposes and BWC treaty. **Obj**: 100% of disclosed BW-related

CBEP Objective	CBEP Capability	Attribute	Measure of Effectiveness (MOE)	Indicator	Evaluation Criteria
Eliminate any BW-related infrastructure and technologies encountered in a partner/country region.	related infrastructure and technologies.				infrastructure is eliminated.

[i] **Collection** is defined as any collections of U.S. Select agents in the same location/facility.

[ii] **Number of locations** must effectively mitigate identified risks based on a threat/risk assessment.

[iii] Confirmation of EDP diagnosis does not require EDP storage except at central repository.

[iv] CBEP refers to **baseline** as no increase in number of EDP collections during pre-engagement.

[v] An **established legal framework** should span all levels (local, regional, national, and international).

[vi] **Biocontainment** is the containment of extremely pathogenic organisms (as viruses) usually by isolation in secure facilities to prevent their accidental release especially during scientific research. http://www.merriam-webster.com/dictionary/biocontainment

[vii] **Epidemic** is the occurrence of more cases of disease than expected in a given area or among a specific group of people over a particular period of time. http://www.cdc.gov/excite/classroom/outbreak/objectives.htm

[viii] BTRP refers to **diagnosis** as the identification and analysis of the nature and underlying cause of an illness or infection. Diagnosis involves laboratory confirmation of the presence or absence of infection and is made by a chemical, microscopic, microbiologic, immunologic, or pathologic study of samples collected from suspect case of infection. For BTRP, diagnosis does not refer to medical diagnosis in which a medical condition or disease is identified/recognized by its outward signs and symptoms.

[ix] Deleted **reporting capability** since not applicable to security.

[x] Metric specifically references **Pandemic Influenza Preparedness plans** because these are the only pandemic plans being created globally. http://www.who.int/csr/disease/influenza/pipguidance2009/en/index.html

[xi] World Health Organization (WHO) References CDC Emergency Preparedness and Response for templates on Bioterrorism Response Planning. http://www.bt.cdc.gov/planning/responseguide.asp

[xii] Multi-hazard IHR hazards of relevance to BTRP include biological (infectious, zoonotic, and foodborne human pathogens). Source: International Health Regulations (IHR) 2005 Monitoring Framework for monitoring Progress in the implementation of IHR Core Capacities in State Parties. Annual Data Collection Tool (2010). http://www.who.int/ihr/checklist/en/index.html

[xiii] "At the national level, state parties are required to assess all reports of urgent events within their territories within 48 hours by applying algorithm contained in Annex II of the IHR." "When a State Party identifies an event as notifiable, it must be notified to

WHO immediately*, i.e. within 24 hrs after having carried out the assessment of public health information related to the event. Such notification will include details of any health measure employed in response to the event as well as accurate and sufficiently detailed public health information available, including case definitions, laboratory results, and number of cases and deaths." "In addition to notification and consultation, state parties are required to inform WHO within 24 hrs of receipt of evidence of public health risks occurring outside their territory that may cause international disease spread." IHR (2005) defines immediately as: within 24 hrs. http://www.who.int/ihr/about/10things/en/index.html

[xiv] **FAO Animal Health Manual** on the preparation of national animal disease emergency preparedness plans. http://www.fao.org/DOCREP/004/X2096E/X2096E00.htm

[xv] **Active surveillance** includes activities in which those responsible for human and animal health identify new cases of disease (case finding) and/or involve interviewing patients, reviewing medical records to detect cases or after index case has been reported. **Passive surveillance** includes activities in which available data on reportable diseases is used or reporting is mandated or requested by those responsible for human and animal health. (Leon Gordis, Epidemiology Second Edition). WHO Recommended Surveillance Standards, Second edition. http://www.who.int/csr/resources/publications/surveillance/whocdscsrisr992.pdf

[xvi] **Events constituting a PHEIC** are assessed by country using the decision instrument in Annex 2 of IHR. http://www.who.int/ihr/9789241596664/en/index.html

[xvii] **OIE reportable diseases** http://www.oie.int/eng/maladies/en_classification2010.htm?e1d7

[xviii] **WHO IHR Monitoring Framework** for monitoring Progress in the implementation of IHR Core Capacities in State Parties. Annual Data Collection Tool (2010). http://www.who.int/ihr/checklist/en/index.html

[xix] **OIE World Animal Health Information System** (WAHIS). http://www.oie.int/ENG/info/en_info.htm

[xx] **Immediate reporting of EDPs** as required by CSTE and referenced by CDC for guidelines regarding nationally notifiable diseases. http://www.cste.org/dnn/LinkClick.aspx?fileticket=7CCxM20JUGg%3d&tabid=36&mid=1496 http://www.who.int/ihr/about/10things/en/index.html

[xxi] **Events constituting a PHEIC** are assessed by country using the decision instrument in Annex 2 of the IHR. http://www.who.int/ihr/9789241596664/en/index.html

[xxii] **OIE reportable diseases** http://www.oie.int/eng/maladies/en_classification2010.htm?e1d7

[xxiii] **WHO/OIE standards** http://www.who.int/ihr/IHR_Monitoring_Framework_Checklist_and_Indicators.pdf

[xxiv] **Optimum conditions** include adherence to appropriate BS&S regulations, cold-chain maintenance, and universal precautions.

[xxv] http://www.who.int/ihr/IHR_Monitoring_Framework_Checklist_and_Indicators.pdf

[xxvi] **Cost** is the combined metric of multiple sustainment topics, including but not limited to: operations, maintenance, technology, infrastructure, knowledge, skills, abilities, educational, logistics, regulatory, cultural) etc. Operations cost is for biosecurity and collection facility. "Maintenance costs" is for biosecurity and collection facility.

APPENDIX C

NATIONAL RESEARCH COUNCIL REPORTS DIRECTLY RELEVANT TO COOPERATIVE THREAT REDUCTION

In addition to the 2009 report *Global Security Engagement: A New Model for Cooperation Threat Reduction*, several National Research Council (NRC) reports have made findings and recommendations directly relevant to metrics for the Cooperative Threat Reduction (CTR) Program key points from those reports are summarized here.

UNITED STATES AGENCY FOR INTERNATIONAL DEVELOPMENT

In 2006, at the request of U.S. Agency for International Development (USAID), the NRC published the report titled *The Fundamental Role of Science and Technology in International Development: An Imperative for the U.S. Agency for International Development.* The report highlights many positive impacts for the United States and developing countries from the application of science and technology (S&T) to international development within U.S. foreign assistance activities. It emphasizes the strengthening of the S&T capacity of developing countries as a key metric in assessing the success of USAID programs, while underscoring self-reliance as the essential component of achieving sustainability. The report discusses three essential components of S&T capacity, namely, (a) institutions that generate, adapt, and disseminate knowledge and technology at international, national, and local levels; (b) technology transfer organizations and individuals that evaluate and adapt knowledge and technologies developed internationally to local conditions; and (c) local institutions that tap and contribute to the world's knowledge supply.

The report stresses the importance of revitalizing the program evaluation capability of the agency. A robust capability to carry out rigorous evaluations of program effectiveness has long been essential to understand the reasons for success or failure of particular programs, according to the report. Also, lessons learned from many types of projects need to be retained, and a key metric in this regard is the number of qualified professionals assigned full-time to the task of program evaluation and who consider such assignments to be career enhancing.

Finally, the report emphasizes the importance of building an S&T culture which spreads to all Missions and all offices of the agency. Other cultures that permeate USAID are the "democracy culture" and the "emergency response" culture. The former has been driven in large measure by congressional earmarks, which over time have required the hiring of 500 USAID advocates who devote their time and skills to promotion of democracy. The latter reflects a widely shared interest at home and abroad, within and outside the agency, in the responsibility of USAID to respond to tragedies throughout the world.

The report argues that there is a widespread latent interest in S&T in all of USAID's partner countries and that expanding staff capabilities from a few dozen to a few hundred would have a dramatic impact in harnessing U.S. S&T in the service of international development. The report sets forth initial personnel hiring targets as an important metric in this regard. It underscores, however, that personnel upgrading must be accompanied by agency leadership, an agency-wide written policy on drawing on the nation's S&T capabilities, available financial resources to prime the pumps of interest in development countries, and an organizational structure that helps ensure that enthusiasm for S&T is captured programmatically.

DEPARTMENT OF ENERGY

Material Protection Control & Accountability Programs

The NRC has carried out three studies supported by the Department of Energy (DOE), on the strengths and weaknesses of the department's cooperative program with Russia on protection, control and accountability of weapons-grade nuclear material (MPC&A program). The reports were as follows: *Proliferation Concerns: Assessing U.S. Efforts to Help Contain Nuclear and Other Dangerous Materials and Technologies in the Former Soviet Union* (1997), *Protecting Nuclear Weapons Materials in Russia* (1999), and *Strengthening Long-Term Nuclear Security: Protecting Weapon-Usable Material in Russia* (2006). Over the lifetime of the MPC&A program, DOE has continuously calculated the quantity of nuclear material of concern that was locked down in secure Russian facilities as its primary metric in judging the success of the program and the impact on non-proliferation. The speed with which the lockdown took place at various facilities was a secondary metric. Eventually, the sustainability of the security upgrades that were installed under the guidance of U.S. experts became a third metric, with sustainability usually measured in terms of Russian financial commitments to supporting and continuing the program and the number of Russian specialists that participated in U.S.-organized training programs.

The first report highlighted the importance of also counting the number of discovered thefts and failed thefts of material as two significant metrics. There were a total of 23 discovered thefts in 1993 and 1994, but zero in 1995 and 1996 from the facilities of the Russian Federation Ministry of Atomic Energy. Also, subsequent reports emphasized the importance of obtaining data on the number of nuclear security violations reported by Russian inspectors of nuclear facilities as indicators of program effectiveness. DOE did not follow up on these recommendations.

Another important recommendation was for DOE department and its Russian counterpart to adopt a ten-year cost-sharing program as a bridge to sustainability. Under this concept, Russia eventually would pick up all costs for the $2 billion program, which had begun with the United States covering all of the costs. While DOE eventually embarked on a sustainability program that rested on cost-sharing, the long-term U.S. financial commitment was much higher than it would have been had the ten-year program been adopted.

Finally, one of the reports offered the following observations:

> There are no good quantitative measures of the effectiveness of U.S. programs to support efforts to upgrade control of sensitive items. First, there is uncertainty as to the effectiveness of existing controls, the security conditions at production and research facilities, and the capabilities of security personnel. In short, there is no good baseline against which to measure progress. Second, it is not possible to separate contributions of the American participation in the program from progress that would have resulted without the Americans. Finally,

there are no reliable data sets concerning legal transfers of sensitive items out of the region, let alone contraband goods, which may not even be known to national authorities.
Reliance must be placed on qualitative assessments as to whether U.S. agencies are effectively using opportunities to upgrade regulatory and security systems. Surrogate indicators–the effectiveness and functioning of components of MPC&A systems and export control systems, for example—are important. Well-developed MPC&A systems in the west provide a basis for comparison in assessing activities in the countries of the former Soviet Union. Considerable attention has been given to opportunities to adapt American experience to different conditions in the target countries. Also, since effective control systems take years to develop, particular attention must be given to interim approaches that can contain leakages in the immediate future.

Protection of Radiation Sources

Hundreds of thousands of ionizing radiation sources, including hundreds and perhaps thousands of highly dangerous sources, were not under adequate control in Russia in the early 2000s. In a 2007 report prepared for the DOE titled *U.S.-Russian Collaboration in Combating Radiological Terrorism*, the NRC identified six key metrics for judging the effectiveness of Russian efforts in reducing both the terrorism and radiological safety aspects of inadequately controlled sources. They were:

- quantification and characterization of existing inventories
- prioritization of source recovery requirements and security enhancements
- enhancement of security at user facilities, during transportation, and at temporary storage sites
- enhancement of final disposition capabilities
- development and implementation of management systems for improved accountability
- development and implementation of research and development priorities in support of the foregoing activities

The report recognized the enormous difficulties and long timelines involved in making measurable progress in reaching each of these goals. At the same time, it urged a nation-wide effort to move forward in all of the areas. Major impediments included:

- decentralized responsibilities in Moscow and throughout the country for undertaking and financing many relevant activities
- chronic shortages of necessary funding, either from the government or from the custodians of sources, to correct security deficiencies
- a legacy of security problems reflected in many inadequately protected radiation sources

Meanwhile, DOE continued its modest efforts to support Russian institutions in estimating the number and locations of sources throughout the country, collecting orphan sources, disposing of unneeded sources, and upgrading disposable areas. Major accomplishments were usually recognized through ribbon-cutting ceremonies. These modest efforts involving U.S. specialists were rapidly magnified through much larger efforts by Russian organizations, and this should have been more widely recognized with DOE as a very impressive accomplishment.

DEPARTMENT OF DEFENSE/DEFENSE THREAT REDUCTION AGENCY

The NRC has published four reports on the Department of Defense (DoD)/Biological Threat Reduction Program (BTRP). Metrics were explicitly emphasized in two of these reports, as indicated below, while many issues directly and indirectly-related to metrics were considered in each of the reports. A few of the relevant observations in the reports are set forth below.

Controlling Dangerous Pathogens: A Blueprint for U.S.-Russia Cooperation (1997)

The report recommended the objectives set forth below as the basis for a long-term cooperative research program, with research activities conducted primarily at sensitive Russian facilities. The measurement of the success in fulfilling research objectives and of the impact of research results was left to others after the research was completed using the initial objectives as a guide in establishing more specific metrics.

- National security benefits: confidence-building, reducing incentives for proliferation, leading to opportunities for dismantlement of biological weapons BW capabilities, enhancing capabilities to combat terrorism
- Public health benefits: improved understanding of characteristics of pathogens that threaten public health; strengthened capabilities to prevent, diagnose, and treat outbreaks of infectious diseases; and enhanced national and international communications concerning disease trends and outbreaks
- Economic benefits: stabilizing Russian institutes, providing opportunities for U.S. private sector investments in Russia, and leveraging limited national financial resources
- Scientific benefits: enhanced knowledge about pathogenesis and increased international availability of research results

Eight pilot research proposals were selected by the NRC authoring committee for financial support by the DoD, according to the following criteria which were to provide the basis for the metrics used to judge success of the specific projects:

- scientific importance of the topic
- quality of the proposal
- quality or capacity of the principal investigator, research team, and facilities
- provision for strong U.S. collaborator
- engagement of former Soviet BW expertise
- promotion of transparency

It was suggested that two additional criteria be introduced as the program expanded. They were (a) the likelihood of sustained support by attracting other sources of finance, and (b) the promotion of linkages between Russian institutions that had been involved in BW activities and those that had not.

Threat Agent Detection Response System Data Base **(2006)**

The report highlighted the following concerns that would complicate both (a) implementing the intended disease surveillance program, and (b) assessing the impact of the program:

- the complex data collection system will challenge many clinicians in host countries who have only limited degrees of computer literacy
- three sets of reporting requirements should be merged into a single system
- stable funding will be difficult to ensure
- training and retention of qualified personnel for the system will be a continuing concern
- governments are apprehensive of U.S. control over all data
- full compliance in providing all data will be difficult
- reviews of large quantities of raw data will produce many false alarms
- effective integration of human and animal disease-related activities will be difficult

The report then concluded that (a) the proposed surveillance network was reasonably well designed on paper, (b) sustainability of the network after DoD completes its efforts to establish the network is critical, and (c) an essential element of sustainability is the broadening of the network from the 16 "low probability" agents, classes of agents, and diseases of interest to DOD to include "high probability agents, classes of agents, and diseases of much greater interest to the host countries. The key metrics measure long-term, not short-term, progress and therefore they must be fully embraced by the host countries.

The Biological Threat Reduction Program of the Department of Defense: From Foreign Assistance to Sustainable Partnerships **(2007)**

This report set forth 54 recommendations for upgrading the BTRP. The overarching theme of the report was the importance of transitioning from a foreign assistance program designed in Washington to a cooperative program designed jointly with partner institutions in the host countries. Five priority recommendations, in addition to the principle of collaboration rather than assistance, addressed the following issues.

- Strong and sustained support for BTRP and related programs is needed.
- The White House should lead the effort to integrate BTRP with related programs of other government departments and agencies, with the authorizing legislation for BTRP requiring involvement of other government agencies.
- BTRP should give greater attention to comprehensive multifaceted engagement for achieving biosecruity while focusing on long-term in addition to short-term payoffs.
- Bioengagement with Russia should be re-invigorated through a variety of cost-shared activities.
- Program management within BTRP needed improvement.

Among the metrics used by BTRP in the past have been the following:

- Number of weapon scientists involved, including the number trained.
- Number of sustainable jobs created.

- Level of matching contributions by cooperating governments or other partners.
- Follow-on contracts resulting from research projects.
- Number of publications in internationally recognized journals.
- Number of patents awarded.
- Number of research products that have reached the market.
- Number of companies that have been spawned.

These indicators were considered important; but they did not address the essence of the program, namely, to what extent has the likelihood of outbreaks of endemic and emerging diseases and the associated terrorist aspects been reduced? Related concerns are the timeliness, adequacy, and quality of responses to outbreaks should they occur. There have been many positive achievements by BTRP at the national and facility levels for addressing infectious diseases. Developing measures for evaluating the changes (e.g., ease of access to sensitive laboratories, response time in identifying outbreaks) could provide the foundation for useful metrics.

***Countering Biological Threats: Challenges for the Department of Defense's Nonproliferation Program Beyond the Former Soviet Union* (2009)**

In the sections focusing on metrics, the report made the following observations:

Inferring the impact of BTRP on a nation's security from laboratory and field evaluations is the most difficult task. Such a task involves understanding the security situation when BTRP entered the scene (the baseline) and the unique contributions of BTRP to reducing biological threats. One approach is for BTRP to support continuing assessments of BTRP impact on risk reduction by both a group of specialists in Washington and a counterpart group of local specialists in the host country. They could develop either common or competing methodologies and then compare results of their assessments. Their different insights as to how BTRP can most effectively enhance security on a broad basis would be of considerable interest.

In short, as BTRP expands into other countries this concern with the impact of BTRP on reducing the threat of bioterrorism deserves greater emphasis. Evaluation efforts should begin from the outset of BTRP involvement in a country. After projects are completed, it will be too late to examine reliable indicators of risk reduction as a result of the projects. At the same time, current midcourse reviews of the Threat Agent Detection and Response [TADR] system in Georgia is a step in the right direction, even though the assumption that TADR is an appropriate approach is not being challenged by the external evaluators—a shortcoming that should be corrected in the future.

The report recommended that BTRP should continue to develop improved metrics that will help guide evaluations of the impact of BTRP, and provide information for setting priorities for activities designed to reduce proliferation of biological weapons as well as related risks from naturally occurring pathogens.

INTERAGENCY

Biological Engagement Programs

The NRC conducted a very broad assessment of U.S. bioengagement activities with Russia. The assessment was published in the report titled *Biological Science and Biotechnology in Russia: Controlling Diseases and Enhancing Security* (2006).

In setting the stage for recommending program approaches and measuring likely impacts, the report identified four U.S. interests in cooperation, namely:

- high potential of Russia's biological research and industrial complex to support both civilian and military programs
- significance of efforts by the Russian government to revitalize research capabilities and apply them to solving social and economic problems
- Russia's vast ecological diversity, which offers unique research environments and provides opportunities for detecting early emergence and movement of dangerous diseases of global importance
- marketing opportunities within Russia for U.S. companies that provide biological products and services

Emphasizing the importance of integrating the Russian health system into global networks that can respond to endemic and emerging diseases, the report stresses the importance of strengthening a broad range of policies and programs of Russia, including the following aspects:

- focus on surveillance, diagnostics, and countermeasures (e.g., drugs and vaccines)
- improve capabilities to detect and diagnose pathogens in both urban and rural environments with improved communication systems
- integrate human and animal disease surveillance
- monitor safety and acceptability of food and water supplies
- strengthen basic research
- strengthen commercialization programs within agriculture and public health
- improve understanding of relationships between infectious agents and non-communicable chronic diseases
- support the biotechnology sector
- strengthen biosecurity procedures at hundreds of facilities
- promote broad transparency
- recruit, train, and retain an expanded cadre of scientists doctors, veterinarians, plant pathologists, and epidemiologists who have access to modern technology to deal with infectious diseases

Then the report focused on five areas where programs should be expanded, results observed, and impact measured:

- improving surveillance and response

- meeting pathogen challenges
- the promise of biotechnology
- the human resource base
- reshaping U.S. Cooperation

In summary, while the report does not explicitly address metrics, it identifies a host of indicators of impact that should lead to approaches in measuring progress toward improved bilateral cooperation that benefits both countries.

APPENDIX D

BIOGRAPHICAL SKETCHES OF COMMITTEE MEMBERS

CHAIR

Jay C. Davis is the President of the Hertz Foundation, which funds graduate studies in the applied physical sciences and engineering. Davis is a nuclear physicist trained at the Universities of Texas and Wisconsin. During his three-decade career at the Lawrence Livermore National Laboratory (LLNL), he built accelerators for research in nuclear physics and for materials science in support of the fusion program. In 1988, Davis founded the Center for Accelerator Mass Spectrometry, the World's most versatile and productive AMS laboratory, creating isotopic tracing and tagging tools for research programs in the geosciences, toxicology, nutritional sciences, oncology, archaeology, and nuclear forensics. At the time he left LLNL to join the Department of Defense (DOD) in 1998, he was the Associate Director for Earth and Environmental Sciences.

MEMBERS

George W. Anderson, Jr. received his Ph.D. in immunology from the Johns Hopkins University. He is a Registered Biosafety Professional with the American Biological Safety Association. Anderson's research experience includes work with Rickettsia, hemorrhagic fever viruses and bacterial agents in high containment laboratories as an U.S. Army, Medical Service Corps Officer. Anderson has been involved with the Cooperative Threat Reduction (CTR) Program for over 10 years, including living in the former Soviet Union while engaged full time in support of the program. He was involved in facility biosafety assessments, biosafety training, laboratory renovation, equipment setup and training, laboratory sustainment operations and preparation for exercises to test the disease surveillance system and diagnostic laboratories engaged by the Defense Threat Reduction Agency in several countries. Anderson is currently the Select Agent Manager at the Lawrence Livermore National Laboratory where he is also the Chair of the Institutional Biosafety Committee and the Institutional Animal Care and Use Committee.

Steven J. Gitomer received his Ph.D. in electrical engineering from the University of Wisconsin-Madison. He joined Los Alamos National Laboratory (LANL) in 1974. He moved to the Los Alamos Center for International Security Affairs in 1995 and was a member of the Nonproliferation and International Security Division beginning in 1993, where his responsibilities included serving as an U.S. member of the Scientific Advisory Committee of the International Science and Technology Center (ISTC), Senior Science Advisor to the U. S. Department of State for the Science and Technology Center in Ukraine (STCU), and principal Los Alamos point-of-contact for the ISTC, STCU, and lab-to-lab interactions with the former Soviet Union. From 1991 to 1993, Gitomer served at the U.S. Department of Energy's (DOE) Office of Arms Control in Washington D.C., where his work focused on implementation of the Threshold Test Ban Treaty and the establishment of the science and technology centers in Russia and Ukraine. Dr. Gitomer retired from LANL in 2005 and became a part-time senior scientist with the U.S. Civilian Research and Development Foundation, he continued his nonproliferation work, adding Iraqi scientist interactions to his portfolio. In 2009, he was appointed as the National Science Foundation program director for plasma physics.

Mary Alice Hayward is Vice President for Strategy, Government and International Affairs at AREVA Inc. in North America. Previously she was Deputy Assistant Secretary for Nuclear Nonproliferation Policy and Negotiations in the Department of State's Bureau of International Security and Nonproliferation. Her portfolio included preventing the smuggling of weapons of mass destruction (WMD) and their delivery systems, implementing international threat reduction programs, developing nuclear nonproliferation policies, tracking, controlling, and securing dangerous materials, including fissile and radiological materials and pathogens, conducting multilateral arms control, nonproliferation, and WMD terrorism negotiations, and developing and shaping nuclear energy policy. Ms. Hayward also served at the U.S. General Accountability Office, the Senate Armed Services Committee, the Department of Energy and the National Security Council where her portfolios focused on reviewing, evaluating and developing policies for the Cooperative Threat Reduction program for the states of the former Soviet Union.

Mark F. Mullen is a project manager in the Nuclear Nonproliferation Division at LANL. He currently serves as the National Technical Director for the Materials Protection, Accounting and the Control (MPC&A) Technology Program, supporting the Department of Energy's Office of Nuclear Energy. He previously served as Assistant Director, Domestic Nuclear Detection Office, Department of Homeland Security, where he led the Systems Architecture directorate. He has more than 30 years of experience in nuclear nonproliferation, WMD threat reduction, homeland security, domestic and international nuclear safeguards and security, and nuclear safety and regulatory issues. Mr. Mullen was deeply involved in early U.S.-Russian cooperative threat reduction programs beginning in 1992, and was one of the principal architects of DOE's laboratory-to-laboratory MPC&A program, which sparked a rapid expansion in U.S.-Russian MPC&A cooperation in the mid-1990s.

Gregory S. Parnell is a professor of systems engineering at the U.S. Military Academy at West Point and is now on sabbatical as a Distinguished Visiting Professor at the U.S. Air Force Academy. His research focuses on decision analysis, risk analysis, resource allocation, and systems engineering for defense; intelligence; homeland security; research and development; and environmental applications. He is Chairman of the Board and a senior principal with Innovative Decisions, Inc., an analytics consulting firm. Parnell is a former president of the Decision Analysis Society of the Institute for Operations Research and Management Science (INFORMS) and of the Military Operations Research Society (MORS). He has also served as editor of *Journal of Military Operations Research.* Parnell has published more than 100 papers and book chapters and has co-edited *Decision Making for Systems Engineering and Management, Wiley Series in Systems Engineering* (2nd Ed, Wiley and Sons, 2011). He has received several professional awards, including the U.S. Army Dr. Wilbur B. Payne Memorial Award for Excellence in Analysis, MORS Clayton Thomas Laureate, two INFORMS Koopman Prizes, and the MORS Rist Prize. He chaired the NRC *Committee on Methodological Improvements to the Department of Homeland Security's Biological Agent Risk Analysis (2008).* He is a fellow of MORS, INFORMS, the International Committee for Systems Engineering, and the Society for Decision Professionals. He received his B.S. in aerospace engineering from the State University of New York at Buffalo, his M.E. in industrial and systems engineering from the University of

Florida, his M.S. in systems management from the University of Southern California, his Ph.D. in engineering-economic systems from Stanford University. Parnell is a retired Air Force Colonel and a graduate of the Industrial College of the Armed Forces.

Kim Kavrell Savit is currently a consulting employee for Science Applications International Corporation (SAIC) and has been an Adjunct Professor at the University of Denver Graduate School of International Studies. Savit retired in May 2006 from her position as the Senior Professional Staff Member for the Middle East, Central and South Asia on the Majority Staff of the United States Senate Foreign Relations Committee. Prior to the SFRC, Savit served in the U.S. State Department as the Deputy Coordinator for Security and Law Enforcement Assistance to Europe and Eurasia (Acting- 2002-2003) and as the Director for Security and Law Enforcement Assistance to the New Independent States of the former Soviet Union (1995-2002). Savit also served in the U.S. Department of Defense as the Director of the Cooperative Threat Reduction Program for the Office of the Secretary of Defense (1991-1995), Country Desk Officer for Morocco, Tunisia, Algeria, Libya, Iran and Iraq, Near East and South Asian Affairs Bureau and as a Budget Analyst for the DoD Comptroller. She received her MA Degree from the Johns Hopkins School of Advanced International Studies. Savit serves as an Advisory Board Member for the Rocky Mountain Region Institute of International Education and as the Chair of the Denver World Affairs Council.

Nicolas van de Walle is a non-resident fellow at the Center for Global Development and is the John S. Knight Professor of International Studies and the Director of the Mario Einaudi Center for International Studies at Cornell University. van de Walle has worked extensively as a consultant for a variety of international and multilateral organizations, including the World Bank, U.S. Agency for International Development, and United Nations Development Programme. His latest book is a Center for Global Development publication Overcoming Stagnation in Aid-Dependent Countries.

APPENDIX E

LIST OF INFORMATION-GATHERING SESSIONS

November 30, 2010
The Board Room, 20 F Street, NW
Washington, DC 20001
AGENDA

9:30 am	Welcoming Remarks Dr. Jay Davis, Committee Chair Background: *Global Security Engagement* *Current committee's mandate and overall study outline*
9:35 am	Briefing from Congressional Staff on Congressional Mandate
10:00 am	Informal introduction from Department of Defense (DoD) CTR Leadership Mr. Ronnie Faircloth, Defense Threat Reduction Agency (DTRA) CTR Director
10:30 am	Briefing from DTRA Program Management Mr. Paul Boren, Director, CTR, DTRA Mr. Leonard Chapman, Deputy Program Manager, Biological Threat Reduction Program Mr. Scott Crow, Program Manager Chemical Weapons Elimination Program Mr. Kevin Sullivan, Weapons of Mass Destruction Proliferation Prevention Program LtCol Bryan Eberhardt, Deputy Program Manager Nuclear Weapons Safety and Security Program LtCol Joseph Kays, Strategic Offensive Arms Elimination Program
11:30 am	Roundtable discussion with DoD Policy, DTRA, Assistant to the Secretary of Defense (Nuclear and Chemical and Biological Defense Programs) Mr. Kenneth Handelman, Acting Assistant Secretary of Defense for Policy Mr. Ronnie Faircloth, DTRA CTR Director
12:30 pm	BREAK
12:45 pm	Working Lunch: Discussion with Guests
2:00 pm	Closing remarks with DoD, including restatement of any action items Dr. Jay Davis, Committee Chair
2:15 am	Public Comment Period
2:30 pm	Adjourn Open Session

April 5, 2011
Keck Center of the National Academies
500 Fifth Street, NW
Washington, DC
AGENDA

Tuesday, April 5

8:30 am	Closed Session
9:00 am	Open Session Welcome, introductions, reminder of statement of task, and outline of goals for the session Dr. Jay Davis, Committee Chair
9:05 am	Lessons learned from exercises and experience implementing new metrics Dr. Beth George, Director, CTR, DTRA Mr. Paul Boren, CAPT Michael Fitzgerald, Mr. Thomas Noon
10:00 am	Discussion
10:40 am	Field visit to Ukraine
11:00 am	Adjourn Open Session

May 17-20, 2011
Committee on Improving Metrics for the Department of Defense Cooperative Threat Reduction Program
Visit to Kiev and Odessa, Ukraine
AGENDA

TUESDAY, MAY 17

Time	*Meeting*	*Meeting with*
1500	SBGSU	General Melnikov (SBSU) and Mr. Simonenko (SCSU)
2000	STCU	Andrew Hood

WEDNESDAY, MAY 18

Time	*Meeting*	*Meeting with*
1000-1130	Central Sanitary Epidemiological Station (CSES)	Dr. Nekrasova and staff
1200 – 1330	State Veterinary Services	Dr. Moroz, Mr. Rublenko, Dr. Golovko and staff
1345		
1500-1630	MinHealth/NSDC	Dr. Onishenko, Deputy Minister and staff
1715	Lunch	
1730	Reception, Defense Attaché's residence	Col. Larm
2154	Overnight train to Odessa	

THURSDAY, MAY 19

Time	*Meeting*	*Meeting with*
1000-1200	Central Reference Lab/Anti-Plague Institute	Tour
1400 – 1530	ICRL, Tour Port of Odessa facility	Maritime SBGU and SCSU

FRIDAY, MAY 20

Time	*Meeting*	*Meeting with*
	Kuchurgan POE	SBGU and SCSU
	Moldovan Border	
	Bolshoi Fontan radar site	SBGU